Maritime Transport and Supply Chain Resilience

Yui-yip Lau • Adolf K. Y. Ng • Zaili Yang
Tianni Wang • Mark Ching-Pong Poo

Maritime Transport and Supply Chain Resilience

Springer

Yui-yip Lau
Division of Business and Hospitality Management, College of Professional and Continuing Education
The Hong Kong Polytechnic University
Hong Kong, China

Adolf K. Y. Ng
International Centre for Resilient Supply Chains, Faculty of Business and Management
Beijing Normal-Hong Kong Baptist University
Zhuhai, China

Zaili Yang
School of Engineering
Liverpool John Moores University
Liverpool, UK

Tianni Wang
College of Transport & Communications
Shanghai Maritime University
Shanghai, China

Mark Ching-Pong Poo
Liverpool Hope Business School
Liverpool Hope University
Liverpool, UK

ISBN 978-3-032-07565-9 ISBN 978-3-032-07566-6 (eBook)
https://doi.org/10.1007/978-3-032-07566-6

This Springer imprint is published by the registered company Springer Nature Switzerland AG
The registered company address is: Gewerbestrasse 11, 6330 Cham, Switzerland

Preface

We can understand resilience as observing, understanding and adapting to constantly occurring global events and enforced/imposed changes very rapidly. In other words, we should become shockproof and not be surprised by any events. However, it appears that while the world is very easily surprised, and more easily shocked, it is very slow to respond and adapt. There is a glaring lack of awareness about the importance of building resilience and robustness within the hard and soft infrastructure of maritime transport and supply chains by way of training manpower to anticipate future potential problems and implement prospective solutions and operating procedures. Neither are there any specially enhanced budgets allocated for human resource development and training on this topic. Maritime transport and supply chains should be resilient equally at the place of origin as well as at destination.

Resilient maritime transport and supply chains could not be built without the active backing and necessary willpower of major political actors, taking initiatives on almost all policies including the development of resilient maritime logistics and global supply chains. But, while focusing on the 'big two' namely the USA and China, we should and could not ignore other major economies that also play significant roles on the global stage in the past but have been diminished since. Taking the cases of Russia, UK, Germany, France, Japan, India, not to say anything of Turkey, Australia, Canada, or Italy. Without the active cooperation from them, it is difficult to imagine how we could reinforce the supply chains and make them more resilient. As the world goes about adopting the fundamental tenets of blue economy to make maritime transport and supply chains more resilient, it highlights the necessity for constant surveillance of trade and allied data to understand trends and patterns developing within the global international trade and deploying of the data so collected to forecast demand for transport and allied services. This will result in automatic matching of supply and demand forces, thus eliminating rapacious and wasteful expenditure and promotion of efficiency and productivity. Building resilience within the supply chains is still in its embryonic stage, given the ongoing shocks that supply chains are constantly facing due to the after-effects of several catastrophic events being played simultaneously on the world stage for the past decade or so.

As such, this book can clarify and highlight the impacts and effects of the several disturbing events that have taken place in the past decade on maritime transport and supply chains—especially at the point of writing, we have just witnessed the second inauguration of Mr. Donald Trump as the new President of the USA, who have promised to continue his economic policies during his first term in office. It is a time of uncertainty, but also a time that all types of opportunities could exist for the betterment of the future.

Last but not least, the publication of this book and its contents would not be possible without the support from different generous sources, namely the National Natural Science Foundation of China (UICR0600041), the Young Scientists Fund of the National Natural Science Foundation of China (Category C) (Funding Number: 52502424), the Research Grants Council of the Hong Kong Special Administrative Region, China (UGC/FDS24/B07/22), CCAPPTIA (www.ccapptia.com), the Canadian Institute for Health Research (CIHR) and Research Manitoba (53013), the European Research Council project under the European Union's Horizon 2020 research innovation program (TRUST CoG 2019 864724) and the International Centre for Resilient Supply Chains, Faculty of Business and Management, Beijing Normal-Hong Kong Baptist University. Also, we would like to thank the publisher, Springer, for the advice and facilitation throughout the publication process. We wish that this book can play a vital part in re-defining the good future of maritime transport and supply chains.

Hong Kong, China — Yui-yip Lau
Zhuhai, China — Adolf K. Y. Ng
Liverpool, UK — Zaili Yang
Shanghai, China — Tianni Wang
Liverpool, UK — Mark Ching-Pong Poo

Acknowledgment

The study was supported by the Research Grants Council of the Hong Kong Special Administrative Region, China (UGC/FDS24/B07/22), National Natural Science Foundation of China (UICR0600041), CCAPPTIA (www.ccapptia.com), the Canadian Institute for Health Research (CIHR), Research Manitoba (53013), the European Research Council project under the European Union's Horizon 2020 research innovation program (TRUST CoG 2019 864724), the Young Scientists Fund of the National Natural Science Foundation of China (Category C) (52502424), and the International Centre for Resilient Supply Chains, Faculty of Business and Management, Beijing Normal-Hong Kong Baptist University.

Contents

The Criticality of Enhancing Resilience for Maritime Transport and Supply Chains

1 Setting the Scene

With the cancellation of sales and shipping contracts and economic contraction, late delivery of goods and emergencies in response to the COVID-19 pandemic, global importers, exporters and the tourism sector have all been significantly affected. Maritime transport and supply chains are no exception. Admittedly, maritime transport and supply chain stakeholders have done a magnificent job in business continuity. Yet, they have been affected by and contributed to a broad range of issues, including: (1) traffic reduction, (2) challenges in conducting operations such as manual paperwork and dock works, (3) hinterland accessibility such as truck driver shortages, (4) the build-up of empty shipping containers and storage of perishable goods and (5) practicing/supporting risk control measures.

In fact, the world nowadays is experiencing different types of uncertainties and risks—some have led to serious disruptions in maritime transport and has affected cruise (e.g. COVID-19 pandemic), container (e.g. skyrocketing of freight rates) and bulk transport (e.g. Russo-Ukrainian and Middle Eastern conflicts). The vulnerability to disruptions were not only due to obsolete systems and facilities. but also short of reliable methods in enhancing maritime transport resilience, such as the lack of scientifically robust data, unreliable models/methods and poor communications/ engagements of vulnerability findings with stakeholders served as the key reasons affecting the quality of resilience planning in maritime transport and supply chains. For example, about climate change impacts, as the maritime sector manages 80% of the international trade, it is considered to contribute a significant part of the GHG emission. In short, maritime transport and supply chains are, part of the issue, and of the solution, and impacted by the climate change. Thus, under a 'business-as-usual' scenario with no new international GHG mitigation measures would pose huge impacts on the global seaborne trade. Also, there are risks that would affect the resilience of maritime transport and supply chain systems, such as vulnerability and

Y.-y. Lau et al., *Maritime Transport and Supply Chain Resilience*,
https://doi.org/10.1007/978-3-032-07566-6_1

security performance, as well as the estimation of safety risk and failure prevention under new technologies (e.g. autonomous ships) and operational environments (e.g. the increasing importance of Arctic shipping) and the impact emission coming from vessel docked at port may have on the inhabitant of port-cities. The emerging risks relating to these new systems and environments would remain hidden until they have been used for a period, requiring the associated risks to be analysed with no or little objective data.

All these are research challenges with theoretical implications that, until now, have yet to be dealt with holistically in the current literature and cannot be solved easily without developing new risk schemes and exchanging knowledge between the domains of climate adaptation, security and safety across the different maritime segments. Also, the wide range of maritime supply chain stakeholders, highly complex port and maritime supply chain operations, and port authorities are important factors for supply chain risk managers to take into consideration. They must consider the interaction among these factors to avoid the silo approach when devising/implementing new measures, especially with unprecedented risks over the past several years which made many supply chain risk managers realise they were not prepared for it, to enhance resilience. This would result in an acceptable level of resilience of maritime transport and supply chains, rather than imbalanced and localised efforts to support the related people, processes and systems. It is challenging to predict the long-term economic effects of the pandemic considering the immense uncertainty. However, exploring the short-term effects, vulnerable actors, activities and systems is imperative.

Before proceeding further, it is important to clarify the concept of resilience. 'Resilience' was conceptually identified by Holling (1973, p. 18) as 'a measure of the ability of ecological systems to absorb changes of state variables, driving variables, and parameters, and still persist'. For many years, engineering practitioners have interpreted resilience as a property of design: its ability to resist a stressor and recover to a steady functional state (Hollnagel et al., 2008). Due to major natural disasters, such as *Hurricane Sandy* in 2012, the concept evolved, and subjects like mitigation methods, standardised assessments and risk-informed management, were added to better describe and evaluate the resilience of systems against future disasters (Linkov et al., 2014). However, the concept has gradually been adjusted to also explain the adaptive capacities of societies, human communities and individuals. In the past decade, resilience has evolved into a notable idea in the glossary of sustainable development and disaster risk management (Tiernan et al., 2019). Also, the rise of high-impact epidemics (e.g. COVID-19) has fundamentally shifted the research direction of resilience from conventional risk-based methods to resilience-based approaches. Fekete et al. (2014) identified science trends, policy trends and paradigm shifts as the reasons for investigating resilience and risk management. Resilience and risk management are urgently demanded because organisations and systems become more associated and complicated, and the effect of incidents becomes greater (Johnsen, 2010). Currently, Blockley et al. (2012, pp. 14) addressed that 'resilience is logically regarded as necessary but not sufficient for sustainability, which implies a stricter requirement needed to achieve the sustainability of a system'. This is crucial to the long-term development of economic and social systems under an associated and changeable international context.

In addition, the very definition of resilience in maritime transport and supply chains has evolved due to the emerging risks in this sector. Previous research on the integration of ports in the supply chain has often been neglected. Ports provide different functions and relate to intermodal transport networks, transformation processes, cargo handling, semi-industrial and industrial functions (e.g. recycling and assembly activities), storage and distribution activities. Ports are regarded as one of the spatial clustering of firms where various maritime supply chain operators participate in creating added value for the final consumers. Therefore, the associations between the focal firm and the port have been identified (Song & Panayides, 2008; Notteboom et al., 2020). Maritime transport and supply chain stakeholders are at a higher level of cooperation and coordination, notably during the pandemic. Also, ports are given a valuable opportunity to change themselves into an integrated and seamless maritime supply chain with a more resilient approach (Ng & Liu, 2014). That said, considering the broad range of port stakeholders, its crucial role in supporting supply chain networks, and consequently its high level of externality, it is imperative to plan and prepare for disruptive events, such as a pandemic and climate change. So now, resilience in ports and the maritime supply chain is not about shutdown or facility damage, but about whether its operation is safe and not hazardous to both people working in ports, the maritime supply chain and society. As such, we must have a thorough understanding of the sector, knowledge of risk and resilience, as well as the intellectual support of experts. The leading disturbance to maritime supply chains is the interruption that happens in global cargo transport created by factors like adverse weather conditions, infrastructure failure and labour shortage. These disruptive incidents have increased uncertainties and logistics costs in various relevant problems. Also, in the past few years, the COVID-19 pandemic has exposed maritime transport and supply chain vulnerabilities in various businesses and threatened safety and security, reinforcing the need for the maritime supply chain to become more agile and resilient (Li et al., 2023).

Based on such a background, we define resilience in maritime transport and supply chains as planning and preparing for changes, and absorbing, recovering from, and adapting to such changes in maritime transport and supply chains. This implies the need to evaluate implemented/potential resilience-building measures (e.g. risk prevention, reduction controls) and determine the resilience level of maritime transport and supply chains. Although systematic research on maritime transport resilience began some years ago, the incorporation of risk models and uncertainty methods into research is still scanty. Hitherto, the research community lacks crucial understanding of maritime risks arising from different threats and uncertainties. This complicates decisions about when, how, and to what extent appropriate strategies and capacity investments should be committed, to successfully adapt to this highly uncertain reality. On the other hand, the bulk of existing research on similar topics on supply chains concentrates on the medical supply chain itself (Golan et al., 2020). Also, most works in infrastructure systems address 'system vulnerability' instead of 'system resilience', such as absorption, recovery, adaptation and transformation (Hosseini & Kash, 2016). In fact, many do not illustrate the complex nature of resilience. In a more comprehensive definition, a resilient infrastructure or organisation's resilience management effectiveness depends on its ability to anticipate, absorb, adapt and quickly recover from a potentially disruptive event (NIAC, 2009). There would be a particular focus on the resilience of maritime transport and supply chains. Leveraging the holistic outlook on the problems, interrelations among people, processes, systems and external events are fully captured, and vulnerable nodes are identified based on their weight in the network. We believe that it is now the right time to encourage more quality research that could cause a paradigm shift on how maritime and supply chain stakeholders should plan for and enact new resilience-based strategies in the future. Although recently there are scholars trying to address this issue—for example, Ng et al. (2025) have edited a special issue on enhancing the resilience of maritime transport in *Transportation Research Part D: Transport and Environment,*[1] where they categorise maritime resilience into four focal areas, namely (1) disruptions and adaptation, (2) environmental concerns and sustainability, (3) the vulnerability of shipping networks, ports and supply chains, and (4) methodological development; this is by no means a complete list, and there is an urgent need for further research on this issue.

Understanding such, this book investigates the state of maritime transport and supply chain resilience by developing emerging topics and real case studies. This offers a better outlook on the resilience of maritime transport and supply chains in emerging markets and mobilises knowledge between them. The intended audiences are researchers, the scientific community, port operations (e.g. port authorities, port operators, shipowners), academic teachers focused on maritime transport and supply chain topics, logistics associations, industrial practitioners, and professionals in the supply chain and related industries. This book can enhance readers' understanding of the recent trends and challenges of maritime transport and supply chain resilience.

2 Contents of the Book

The rest of the chapter briefly introduces the contents of the book. Chapter 2 dissects the climate adaptation strategies and planning of maritime transport and supply chains. The study of maritime resilience primarily has its roots on the attention to the mitigation and adaptation to climate change impacts to maritime transport and supply chains. Indeed, maritime transport and supply chains are exposed to the vulnerability of climate change, mainly because of their location, including low-lying areas, coastal zones and deltas. While there is increasing research on climate adaptation strategies and planning of ports, hitherto, there is a lack of works that

[1] For further information, visit TRD's website: https://www.sciencedirect.com/special-issue/10CV3TC3G8S.

explain how scholars address the topic from different theoretical angles. The chapter conducts the discussion from four theoretical angles, namely institutional system, path dependence, supply chain risk management and stakeholder management. It is a germane reminder to stakeholders that effective climate adaptation is not limited to engineering technicalities but is an ideological issue that requires a fundamental shift of the existing political, economic and social paradigms.

Chapter 3 offers methodological contributions on the topic through conducting uncertainty modelling on climate risk. Maritime transportation makes significant contribution to economic prosperity and suffers much from various risks, including classical (e.g. ship collisions) and emerging ones such as those relating to climate change risks (e.g. flooding, heatwaves, storms). Data uncertainty represents an important challenge for safety science in general climate risk analysis in specific given the insufficiency of historical failure data, compared to classical risks. Broadly data uncertainty is categorised into three groups—fuzziness, incompleteness and randomness—hence triggers risk studies using the uncertainty theories such as fuzzy logic, Dempster-Shafer (D-S) theory of evidence and Bayesian probabilistic inference. This chapter's aim is to introduce the newest study on the development of climate risk modelling using the three uncertainty theories, individually and collectively within the maritime transport context. It will help understand the state of the art of the relevant research and inspire new ideas.

Chapter 4 delves into the central position and resilience of China's major ports in the global shipping network under the influence of typhoons. This study focuses on the ports of Shanghai, Shenzhen, Qingdao, Xiamen and Qinzhou, and analyses their network centrality and resilience through a comprehensive evaluation framework. In terms of centrality evaluation, a shipping network was constructed using UCINET software, and two new dimensions, feature vector centrality and core edge analysis, were introduced based on conventional evaluation dimensions to improve the accuracy and completeness of analysing the importance of ports in the shipping network. In terms of resilience assessment, a port resilience evaluation index system was constructed, and research methods such as Analytic Hierarchy Process (AHP), Entropy Weight Method, and TOPSIS Method were used to evaluate the port's ability to resist typhoon interference. Research has shown that the five sample ports rank high in centrality within the constructed shipping network. Among them, Shanghai Port maintains a dominant position in global shipping and is crucial for global trade and regional economic prosperity.

Chapter 5 discusses the building of resilient maritime transport and supply chain networks in the Yangtze River Delta in China. With the intensification of global climate change and the frequent occurrence of natural disasters, the resilience of ports, as crucial nodes in the global supply chain, has become extremely important. The Yangtze River Delta is one of the most economically active regions in China, and ports such as Shanghai Port and Ningbo Zhoushan Port play significant roles in trade. However, in recent years, the frequent occurrence of extreme weather events has brought great challenges and numerous risks to the port trade in the Yangtze River Delta region, which also indirectly affects the stable operation of the global supply chain. Therefore, this paper takes the coastal ports in the Yangtze River Delta as the research object, adopts the risk assessment method based on an index system, and applies the matter-element extension risk assessment model to analyse the impacts of two representative extreme meteorological disasters, namely tropical cyclones and storm surges, on the infrastructure and operation of coastal ports in the Yangtze River Delta region, quantify the risk levels of extreme meteorological disasters in ports, analyse the influence degree of each area of the ports when suffering from extreme meteorological disasters, and propose strategies to improve resilience based on the results, providing support for ensuring the safe operation of the supply chain.

Chapter 6 discusses the building of resilient maritime transport and supply chain networks among ASEAN countries in Southeast Asia. ASEAN countries' maritime transport and supply chain networks are critical to global trade and tourism, yet they face increasing challenges from climate change, geopolitical shifts, and pandemic-induced disruptions. This chapter focuses on rebuilding resilience in ASEAN's maritime systems by examining key ports such as Singapore, Port Klang, and Ho Chi Minh City, which play diverse roles within the region's interconnected network. Using a case study on the ASEAN cruise port network, the chapter introduces a vulnerability assessment framework that leverages social network analysis (SNA) and centrality measures, augmented with a COVID-19 Performance Index (CPI), to evaluate port-level and network-level impacts of local disruptions. Comparative analyses of pre- and post-pandemic periods reveal shifts in port rankings and network efficiency, highlighting the resilience of major hubs like Singapore while exposing vulnerabilities in smaller ports. The chapter concludes with actionable policy recommendations, including climate-adaptive infrastructure, digitalisation and regional collaboration, to enhance the resilience of ASEAN's maritime transport and supply chain networks in the face of global uncertainties. The findings contribute to the discourse on sustainable maritime systems and inform strategies for fostering resilience in similar regions globally.

Chapter 7 explores how cloud-based blockchain technology for vaccine supply chain applications can be technically sustainable in the Greater Bay Area. To explore the technical sustainability, the user's intention to adopt the cloud-based blockchain for vaccine supply chain applications has to be proposed, especially at this infancy stage of implementing the supply chain applications using such a new blockchain technology. For this exploration of technical sustainability, the integration of the models of Task-Technology Fit and Unified Theory of Acceptance and Use of Technology is used. The significance of the book chapter is twofold. First, there is a trend of adopting this new blockchain technology in supply chains around the world for the main reason that the blockchain features present promising opportunities for improvement across the supply chain operations around the world. This proposed study investigates this important issue of sustainable development of blockchain technology for vaccine supply chain applications in the Greater Bay Area in order to understand the feasibility of adopting blockchain technology in supply chains in the Greater Bay Area to cope with the trend in the world. For this investigation, a new and more appropriate model by integrating Task-Technology Fit and the Unified Theory of Acceptance and Use of Technology is adopted. Second, while previous studies suggested and reported sustainable practices of applying blockchain technology in supply chains in many countries around the world, how these practices apply to the Greater Bay Area has not been thoroughly explored in the literature. This proposed study targets to fill this gap by exploring technical sustainable practices.

Chapter 8 discusses the complexity of adaptation strategies for maritime transport. Maritime transport systems face increasing complexity in adapting to global challenges, including climate change, pandemics and geopolitical tensions like war. It examines the adaptation strategies of UK ports, focusing on specific climate risks, including rising sea levels, extreme weather, and shifting trade patterns, and corresponding resilience measures, such as coastal defences, green technologies, and flexible port management. It also explores public responses to logistical needs during global disruptions like the COVID-19 pandemic and war outbreaks, highlighting evolving demands for essential supplies and reconstruction logistics. This chapter addresses the complexities of climate and global events and provides actionable insights to sustain maritime operations and supply chains in an increasingly uncertain world.

Chapter 9 addresses e-commerce and resilient supply chain management during the COVID-19 pandemic. Effective mitigation of pandemic challenges, such as COVID-19, is a key issue that impacts human lives. Urban, economic and social activities involve supply chains, meaning that developing resilient supply chains is essential for prosperity. There is a scarcity of knowledge on how the supply chain system can adapt to large-scale pandemic outbreaks under a lockdown policy. Focusing on COVID-19, the book chapter aims to develop effective supply chain planning and solutions against pandemic outbreaks. The key objectives are therefore: (1) to investigate how individuals with different characteristics (e.g. age, locations, jobs) perceive and react to lock-down policy in tackling the pandemic, (2) to investigate how the general public responds to the different logistical needs in pandemic outbursts at different stages; (3) to identify ways that can facilitate the transfer of lock-down policies to cities and countries; and (4) to develop policies and a supporting framework for pandemic supply chain planning under lock-down policy.

Chapter 10 provides a historical case on the impacts of a pandemic on the resilience of supply chains and economies, namely the case of COVID-19 in Manitoba, Canada. The effects of COVID-19 on the for-profit sectors in Canada have been remarkable, as evidenced by the increasing number of businesses filing cases for bankruptcy and high unemployment rates. Among the Canadian provinces, Manitoba has the highest rate of COVID-19 confirmed cases across Canada. As such, the chapter investigates the key aspects of Manitoba, including Canada's and Manitoba's COVID-19 pandemic-related laws, panic buying and border closure with the United States and other Canadian provinces during the pandemic era. Some lessons on possible strategies and opportunities to revitalise Manitoba, and indeed other similar cases in the future, during the post pandemic era, are provided, including enhancing the resilience of supply chains. The Manitoban experience could be a valuable illustrative exposition for other regions in handling similar crisis.

Finally, based on the discussions in previous chapters, Chap. 11 provides the best practices and guidelines for resilient maritime transport and supply chains. It synthesises successful stories and practical insights for designing and implementing resilient maritime transport and supply chains in an era defined by climate change, geopolitical uncertainties and post-pandemic recovery. Also, it highlights strategies that have successfully enhanced operational robustness, reduced vulnerabilities and sustained global trade flows by analysing real-world examples. It introduces an integral resilience model to guide the development of new systems, winning strategies and informed decision-making for resilience building. These frameworks prioritise adaptability, collaboration and innovation, offering actionable guidelines for stakeholders to navigate the challenges of a dynamic global environment. Finally, it concludes with a comprehensive book summary, consolidating its key findings and providing a cohesive roadmap for advancing sustainable and resilient maritime systems worldwide.

3 Previously Published Materials

Some of this book's contents have previously been published in other forms of publications. They include:

- Ng, A.K.Y., Wang, T., Poo, C.P., Becker, A. and Lau, Y.Y. (2023): 'Dissecting climate adaptation strategies and planning of ports from different theoretical angles'. *Paper presented at the Annual Conference of the World Transport Convention (WTC) 2023*, Wuhan, China, 14–17 June 2023.
- Poo, M. C. P., Yang, Z., & Lau, Y. Y. (2024). Vulnerability analysis of cruise shipping in ASEAN countries facing COVID-19 pandemic. Ocean & Coastal Management, 248, 106919.

References

Blockley, D., Godfrey, P., & Agarwal, J. (2012). Infrastructure resilience for high-impact low-chance risks. *Proceedings of the ICE—Civil Engineering, 165*(CE6), 13–19.

Fekete, A., Gabriele, H., & Sylvia, K. (2014). Benefits and challenges of resilience and vulnerability for disaster risk management. *International Journal of Disaster Risk Science, 5*(1), 3–20.

Golan, M. S., Jernegan, L. H., & Linkov, I. (2020). Trends and applications of resilience analytics in supply chain modelling: Systematic literature review in the context of the COVID-19 pandemic. *Environment Systems and Decisions, 40*, 222–243.

Holling, C. S. (1973). Resilience and stability of ecological systems. *Annual Review of Ecology and Systematics, 4*(1), 1–23.

Hollnagel, E., Nemeth, C. P., & Dekker, S. (2008). *Resilience engineering perspectives: Remaining sensitive to the possibility of failure*. Ashgate.

Hosseini, S., & Kash, B. (2016). Modeling infrastructure resilience using Bayesian networks: A case study of inland waterway ports. *Computers and Industrial Engineering, 93*, 252–266.

Johnsen, S. (2010). Resilience in risk analysis and risk assessment. In T. Moore & S. Shenoi (Eds.), *Critical infrastructure protection IV* (Vol. 342, pp. 2015–2027). IFIP AICT.

Li, D., Jiao, J., Wang, S., & Zhou, G. (2023). Supply chain resilience from the maritime transportation perspective: A bibliometric analysis and research directions. *Fundamental Research.* https://doi.org/10.1016/j.fmre.2023.04.003.

Linkov, I., Bridges, T., Creutzig, F., Decker, J., Fox-Lent, C., Kröger, W., Lambert, J. H., Levermann, A., Montreuil, B., & Nathwani, J. (2014). Changing the resilience paradigm. *Nature Climate Change, 4*(6), 407–409.

Ng, A. K. Y., & Liu, J. J. (2014). *Port-focal logistics and global supply chains*. Palgrave Macmillan.

Ng, A. K. Y., Faury, O., Lau, Y. Y., Becker, A., & Munim, Z. H. (2025). Enhancing the resilience of maritime transport: theory and practice. *Transportation Research Part D: Transport and Environment* (in press).

NIAC. (2009). *Critical Infrastructure resilience final report and recommendations*. www.dhs.gov/xlibrary/assets/.../niac_critical_infrastructure_resilience.pdf

Notteboom, T., Van Der Lugt, L., Van Saase, N., Sel, S., & Neyens, K. (2020). The role of seaports in green supply chain management: Initiatives, attitudes, and perspectives in Rotterdoom, Antwerp, North Sea Port, and Zeebrugge. *Sustainability, 12*, 1688–1711.

Song, D. W., & Panayides, P. M. (2008). Global supply chain and port/terminal: Integration and competitiveness. *Maritime Policy & Management, 35*(1), 73–87.

Tiernan, A., Drennan, L., Nalau, J., Onyango, E., Morrissey, L., & Mackey, B. (2019). A review of themes in disaster resilience literature and international practice since 2012. *Policy Design and Practice, 2*(1), 53–74.

Theoretical Discussions on the Climate Adaptation Strategies and Planning of Maritime Transport and Supply Chains

1 Introduction

The study of maritime resilience primarily has its roots on the attention to the mitigation and adaptation to climate change impacts to maritime transport and supply chains. Since the last century, climate change has been at the forefront due to its potential catastrophic risks posed to human lives and activities (UNECE, 2010; Intergovernmental Panel on Climate Change, 2012). By now, it is probably too late to avoid all the deleterious effects of climate change, in no small part due to uncertainties on how the problem should be addressed (Applegate, 2010). Thus, adaptation is not a choice but a necessity, including transport infrastructures and facilities. For example, located along shorelines, ports are especially vulnerable to the risks that climate change can pose to their facilities and operations (Becker et al., 2012). Despite evidences suggesting that institutional systems will influence climate change adaptation due to ambiguity and competing political interests (Wheeler et al., 2009; Keohane & Victor, 2010; Preston et al., 2011; Osthorst & Manz, 2012), with UNCTAD (2012) calling for both 'hard' and 'soft' strategies to develop resilience to climate change impacts, most attention focuses on physical layouts and technical details of capital-intensive engineering projects, such as elevation, levee, dykes, to name but a few (National Research Council (NRC), 2010).

The awareness of climate change is growing, as their consequences become increasingly apparent. 40% of the global population lives within 100 km of the coast, and port cities are significant concentrations, where the coasts are often intensely populated (such as major port cities). Extreme weather events, supercharged by climate change, affected more than 60 million people around the world (United Nations, 2019). Highly populated port cities are in areas vulnerable to climate change impacts: on coasts susceptible to sea-level rise and storms or at mouths of rivers susceptible to flooding (Becker et al., 2012). In the past few years, there has been a growing interest among researchers and practitioners to reduce the carbon

Y.-y. Lau et al., *Maritime Transport and Supply Chain Resilience*,
https://doi.org/10.1007/978-3-032-07566-6_2

footprint of maritime shipping and aviation logistics for mitigating climate change effect by adopting operations management practices. These include operational decisions such as speed reduction, berth scheduling and route re-engineering to rationalise fuel consumption and to reduce CO_2 emissions. On the adaptation direction, there is growing interest, but mainly on climate vulnerability assessments and risk assessments (Poo et al., 2018) and yet to implement operations management practices. With increasingly frequent, severe climate-related events, adapting to the impacts posed by climate change has recently become a pivotal research topic influencing transport operation, infrastructure, planning and policymaking. For instance, it urgently requires illustrating the *status quo* regarding long-term risks posed by climate change on ports, including detailed analyses of the current measures and dilemmas in handling the issues of climate change and adaptation of planning to provide competent advice with port stakeholders. Hitherto, most research on climate adaptation focus on the short-term impacts, and there is insufficient research on how to systematically adapt to the effects of climate change on ports, especially in the reduction of uncertainties in decision-making, the development of effective public policies, and best practices. Of course, this is not to deny the increasing contributions of scholars in conducting research on the topic trying to explain the strategic and planning process from different angles, including the authors of this chapter (e.g. Becker et al., 2013; Xiao et al., 2015; Ng et al., 2016; Becker et al., 2018; Ng et al., 2018a, b; Yang et al., 2018; Ng et al., 2019; Jiang et al., 2020; Lin et al., 2020; Panahi et al., 2020). Nevertheless, it is somewhat surprising that, to our best knowledge, there is a lack of articles that explain how scholars dissect climate adaptation strategies and planning of maritime transport and supply chains. This chapter aims to fill in this gap.

2 Institutional System

An institutional system is a set of mechanisms, establishments and practices that structure relationships between agents, both public and private. It imposes preceding constraints on policymaking choices and strategic directions (March & Olsen, 1989; Steinmo et al., 1992; Hall & Taylor, 1998). It countervails dramatic changes, restricts alternatives and diminishes the rationalities of decision-making to predictable paths according to norms and practices based on culture and hegemonic values of the time (Fuchs & Scharmanski, 2009; Glassman, 2004) even when they may have become obsolete (North, 1990; Hodgson, 1993). Institutional systems solidify generally accepted values of a society into predictable practices to deter (or at least control) undesirable social outcomes due to individual actions. Having said so, it could stretch (Strambach, 2010; Notteboom et al., 2013) with changing circumstances. The stretching involves two major components, namely the institutional environment and the institutional arrangement. The former refers to informal conventions and norms of which organisations, being parts of a given community, should conform to gain legitimacy and general support, and sometimes made compulsory through legally binding rules and regulations (Martin, 2000). The institutional environment forms the basis for compromise (Gutmann & Thompson, 2012), operational characteristics and receptiveness to new knowledge (Boxer, 1991). The institutional arrangement refers to agreements and organisational structures between agents to achieve certain objectives or programs governed by the institutional environment, like firms, bureaucracy, policies and cooperative networks.

The influence of the institutional system on maritime transport has been widely studied (e.g. Buitelaar et al., 2007; Ng & Pallis, 2010; Notteboom et al., 2013). However, they largely followed the neo-institutional approach that investigated how established institutional environments structure cognition and guide decision-making. How and why the institutional system mattered has remained largely untouched. Facing circumstances like climate change, institutional agents might take spontaneous initiatives to re-structure the institutional arrangements, as exemplified by the neoliberal institutional and management reforms among maritime transport and supply chain systems, such as ports. However, decision-making got more complex within an uncertain institutional environment consisting of individual mindsets, ambivalent interests and diversified localities with individualistic and pluralistic traditions (Fishman, 2000). Climate adaptation planning possessed such an uncertain environment due to scarce legal standards, direct precedents and readily transferrable scientific knowledge. This caused inadequate understanding, and thus inadequate input, from stakeholders and the public. Planning should provide clear guidance and practical actions to lead the direction of development, especially in the generation of first plans with many (untried) alternatives to choose from (cf. Wheeler, 2008; Preston et al., 2011; Sager, 2011). Further problems arise when the new circumstance has yet to reach a critical juncture (Buitelaar et al., 2007) and all stakeholders do not yet deem significant transformation necessary or immediate.

3 Path Dependency

Path dependence theory posits that prior conditions and events constrain and shape later outcomes and policy options (Van Driel and Devos, 2007, pp. 682). It emphasises the uncertainties in a process that the original form of phenomenon is impacted by initial conditions and the development of subsequent form restricted by lock-in effects (Van Driel and Devos, 2007, pp. 682). This concept has been widely applied to port studies in diverse ways (Notteboom et al., 2013). In the case of ports, for example, there are three streams that deal with its evolution. First, it is the long-term evolution of port systems that focuses on the changing relationship between the port and the port city (Norcliffe et al., 1996). Second, it is the governance systems of ports (Brooks and Cullinane, 2007) that treats ports as agglomerations of relevant industries and concerned with the unique development trajectories of ports and the diversity of management structures. Third, it emphasises the role of path dependence in institutional economics and place that implies the differences of institutional path dependence with the change in ways and places. As per Notteboom et al. (2013), all the stated streams involve dynamic interactions among institutional environment and government structure. With the impacts and uncertainties of climate change being increasingly serious and unpredictable, the development and evolution of maritime and supply chain systems require stakeholders to consider the dynamics and stability in the governance system and geographical conditions. Hence, it is not difficult to understand that a climate adaptation plan (usually through considering the potential risks and uncertainties) is a positive response to the path dependence theory. Also, every path dependence system has evolved thanks to different histories of their own systems and owing to the disproportionate impacts of climate change in different geographical areas. In this case, maritime and supply chain stakeholders need to consider the specific conditions at every individual port. It is a chronological process that decisions at an initial stage would affect the consequence of events in latter stages (Pettigrew, 1992). This encourages stakeholders to consider the sequences of events' occurrence and improve the elements in climate change when making an adaptation plan for maritime transport and supply chains.

Meanwhile, path dependence theory has been applied in studies of long-term strategic planning (Dooms et al., 2013), albeit in a more limited way. Pierson (2000) and Kay (2005) suggested that the path dependence concept is related to analysis of temporal dynamics, which emphasises that significant changes in investment strategy and governance are often attributed to the shift of exogenous events (e.g. climate change impacts) and the role of stakeholders. Since decision-making is a multi-party involved process, path dependence indicates that institutions in the path could be substituted when the existing institutions and conditions cannot approach the objective. Therefore, a long-term, strategic plan based on 'real stakeholder inclusion' can play key roles in governance change and development (Dooms et al., 2013). In other words, to establish a resilient framework for ports' adaptation planning to minimise climate change risks and impacts, policymakers need to consider the uncertainties of climate change and its impacts.

4 Supply Chain Risk Management (SCRM)

With supply chain management and risk management, each having been established and developed, their intersection area, supply chain risk management (SCRM), the definition attention in the past decades (Paulsson, 2004). In the context of SCRM, risk can be regarded as 'unreliable and uncertain resources resulting in supply chain interruption', while uncertainty can be interpreted as 'matching risk between supply and demand in supply chain processes' (Tang and Musa, 2011, pp. 26). Here we acknowledge that both risks and uncertainties in SCRM are often connected to negative consequences and are difficult to distinguish in most literatures (e.g. Wagner and Bode, 2006).

There are considerable studies investigating the type of supply chain risks. Generally, supply chain risks contain supply breakdowns, disruption, purchasing and forecasting failures (Johnson, 2001; Zsidisin, 2003). As a major risk, supply chain disruption can be divided into unintentional and intentional disruptions. Unintentional disruption includes natural disasters (e.g. hurricanes, tornados, and floods that disrupt supply paths, transportation and manufacturing infrastructures) and man-made accidents (e.g. transportation-related injuries that lead to delays, and further negative effects on products and services). In this regard, Tang and Musa (2011) categorised the risks in supply chain by material flow risks, financial flow risks and information flow risks in a literature survey from 1995 to 2008. The risks in material flow stem from source, make and delivery aspects, and represented by variables in sourcing flexibility, suppler selection, product monitoring, product process and design, and demand seasonality (Fitzgerald, 2005; Christopher and Peck, 2004; Khan et al., 2008; Peck, 2005; Bovet, 2006). In financial flows, the risks come from unstable factors in exchange rate, price and cost, financial strength of supply chain partners, and financial handling and practice (i.e. Johnson, 2001; Bovet, 2006; Hartley-Urquhart, 2006). Also, in information flow, the risks include information accuracy, information system security and disruption (i.e. Lee, 2004; Finch, 2004).

Meanwhile, with the increasing changes in the market and the long-term consideration of strategic decisions in supply chain network design (Baghalian et al., 2013), uncertainty plays a more prominent role in supply chain management. Hitherto, there is attention on lower-impact and unintentional risks that involves the risks and uncertainties posed by, to some extent, climate change in supply chains. SCRM, as a major part of SC management, requires designing a dynamic supply chain network structure by considering both risks and uncertainties to manage the product flow throughout the system to predict, arrange and recover from disruptions (Baghalian et al., 2013). Therefore, a long-term adaptation plan for climate change could minimise the risks and uncertainties and increase the dynamics in SCRM. In this case, a literature survey on SCRM from 1995 to 2009 by Tang and Musa (2011) demonstrated a growing volume of publications since 2004 from both academia and industries. At the same time, frequent discussion topics include challenges and opportunities of outsourcing to low-cost countries, information security and sharing, partner relationship, economy, environmental and political issues in supply chain, etc. Nevertheless, there are some gaps in SCRM research. Although some efforts have been extended to combine material and cash flows, most of the literature still relies on the issues in material flow, especially in supplier selection. Also, even though enough awareness has been given to SCRM in industry, the focus still relies on qualitative aspects (e.g. descriptive model, conceptual model) and there is a lack of quantitative research (e.g. quantitative model, risk-related information). Based on this, potential approaches in developing quantitative models have been proposed, such as robust planning, system dynamics and reverse logistics. This enlightens us to establish an adaptation plan for risks and uncertainties posed by climate change by combining qualitative and quantitative methods.

Here it is noted that while there is a widespread recognition that a port is a key node in a supply chain, and supply chain management should be applied to the port sector, port risk management has not been given sufficient attention. Recent research in supply chain risk management tends to focus on identification of disruptions and mitigation countermeasure for catastrophic events (Guerrero et al., 2008). Meanwhile, researchers tend to focus on risk factors in the change of demand and marketplace in supply chains (e.g. Das, 2011). However, given the fragility and complexity of supply chains, adaption planning by considering uncertainties in climate change should be applied in risk management.

5 Stakeholder Management

The theory on stakeholder management is originated from the field of strategy, where stakeholder is defined as 'any group or individual who can affect or is affected by the achievement of the organization's objective' (Freemen, 1984, pp. 53). In this case, the stakeholder theory is 'managerial and recommends the attitudes, structures and practices that, taken together constitute a stakeholder management philosophy' (Donaldson and Preston, 1995, pp. 67).

Smudde and Courtright (2011) conducted holistic research in stakeholder management based on three primary questions: '(1) how are stakeholders created? (2) how can relationships with stakeholders be maintained? (3) how can relationships with stakeholders be improved?' (pp. 137). Meanwhile, in response to the demand of strategic planning, two realms of stakeholder management have been elaborated in Smudde and Courtright (2011)'s article to enhance the dynamics of this field: reactive and proactive stakeholder management. Specifically, reactive stakeholder management focuses on past activities that have affected relationships with stakeholders. Through a single loop learning process, reactive stakeholder management highlights the need to draw past experiences including what went right and wrong, what deficiencies were eliminated or reduced and what could be improved in the future, etc. Proactive stakeholder management is concerned with future activities to produce opportunities for cooperation between any stakeholders and the organisation. The proactive concept involves a double-loop learning process where an organisation is adaptively adjusting to the management process in a changeable environment. The advantage of proactive stakeholder management relies on that, via an adaptive and tactics design, it would maximise (or enhance) the strengths, minimise the weakness, generate opportunities and screen threats for success in the future. Accordingly, climate adaptation strategies and planning of ports is consistent with proactive stakeholder management theory in predicting the potential damages, crises and disasters, managing the relevant personals who are and will be stakeholders and improving the stakeholder relationships.

Fassin (2012) analysed strategy management, reciprocity and responsibility. Stakeholders can be broken down to the four categories in terms of their legitimacy: stakeowners, who are the internal constituents having a real stake in the company; stakewatchers, who are the mainly pressure groups influencing the firm; stakekeepers, who are the mainly regulators imposing external control and regulations on the firm; and stakeseekers, who are seeking a voice in a corporation's decision-making

(Holzer, 2008). While stakeholder theory has mainly focused on corporate responsibility toward a firm's stakeholders (Friedman and Miles, 2006; Freeman et al., 2007) from the strategic perspective, Fassin (2012) proposed that stakeholders could also influence the corporation (stakeholder reciprocity and responsibility). More specifically, the stakeholder reciprocity could be limited by the stakeowners who legislated stakes and loyal partners are involving multiple benefits; meanwhile, the stakeholder in stakeholder responsibility refers to the corporate stakeholders excluded from the stakeowners. Long-term commitment and responsibility require more attention to the reciprocity from stakeowners and their moral responsibility including loyalty, fairness and ethical treatment. Also, stakewatchers and stakeseekers formulated their strategy from the political resource's perspective; this should not underestimate their ethical responsibility. Applied to maritime transport and supply chains, stakeholders are encouraged to recognise the diverse roles in stakeholder's networks and be concerned with the reciprocity of stakeonwers and responsibility of different stakeholders when making an adaptation plan.

Existing literature has, to date, provided applicable guidance for stakeholder management in governance and planning. First, stakeholder management reflects the demands of increasingly intensive competition and integration of supply chains (Lam et al., 2013). For example, a 'port-focal' supply chain (cf. Ng & Liu, 2014) implies that the port as its core enterprise, combines and coordinates service providers with customers into an effective system that distributes the correct number of goods to the right place, at the right time, and to maximise the efficiency and profits of the entire supply chain (Cheng and Zhu, 2011, pp. 1). This requires a comprehensive stakeholder management process, where stakeholders emphasise the roles, responsibilities, and relevance of stakeholders, and consider internal and external uncertainties in port-focal supply chains. As Dooms et al. (2013) have pointed out, the divergent preferences of stakeholders for the desirability of further port development, how to balance the multiple benefits of port stakeholders in supply chain is another question that must be thoroughly considered. Therefore, a long-term, strategic plan based on stakeholder management would contribute to change in the maritime transport and supply chain systems.

Second, stakeholder management contributes to regional governance. For example, in ports, stakeholders, involving shipping companies, shippers, terminal operators, government, logistics service providers and other parties associated with port industries are expected to create and sustain value in the same customer value chain (Debrie et al., 2013; Lam et al., 2013). In particular, the regional governance broadens the scope of stakeholders to policymakers at diverse levels, community groups and market players at diverse ports (Preston and Sapienza, 1991). The complexity of stakeholder management in regional governance calls for the collaboration of multiple parties through joint projects and technological innovation which is supported by a strategic port planning of policymakers, to minimise the conflicts among stakeholders and maximise their common benefits (Lam et al., 2013). This theory can be also applied into adaptation planning. In response to the increasing risks and uncertainties of climate change on maritime infrastructures and facilities in a wider area, planning is expected to accommodate the change of the external environment, encourage an extensive participation of stakeholders, reduce the risks and conflict among stakeholders, create a technology framework and tailor to the specific port conditions by a joint effort against to climate change.

6 Conclusion

Recognising the lack of articles that explain how scholars dissect climate adaptation strategies and planning, this chapter conducts a theoretical discussion on the climate adaptation strategies and planning of maritime transport and supply chains. It looks at the topic from four theoretical perspectives, including institutional system, path dependence, supply chain risk management (SCRM) and stakeholder management.

Institutional system plays a significant role in affecting the process of climate adaptation planning, as well as its effective implementation. Hence, without doubt, it would be a far-reaching attempt for any stakeholders to address climate change adaptation. By doing so, it would trigger some re-structuring within the institutional arrangements and demonstrated planners' awareness that paradigm shift from previous planning norms and practices were necessary. However, uncertainty in the institutional environment and the likely speculative attitude of major participants would strengthen the perception that political controversies would hinder future implementation. Without resolving these challenges, climate adaptation plans might become more of a visionary guide rather than a real action plan, and planners might be forced to muddle through the process by undertaking a highly evolutionary approach. Perhaps it is not surprising, as the objective of the institutional system was to deter undesired shocks to societies due to individual actions (Weber, 1922). Under such an uncertain institutional environment, the neoliberal ideology, which emphasises minimal public intervention (Harvey, 2005), might continue to dominate planning decisions. Under such influence, planners might be forced to adopt an evolutionary approach in climate adaptation planning, even if they favour a more revolutionary approach.

On the other hand, path dependence emphasises the significant changes in investment strategy and governance are often attributed to the shift of exogenous events (e.g. the climate change impacts) and role of stakeholders. For example, some ports in developed countries have realised the impacts of climate change occurred in the past and actively adapt to these impacts. Meanwhile, the motivations of an adaptation plan are consistent with the two critical factors in path dependence: the historical climate change event or small changes, and the prediction for climate change in the future. However, several climate change uncertainties are inadequately considered when predicting the climate change in the future. Path dependence theory requires planners to establish a resilient framework to minimised climate change risks and uncertainties as well as concerned with the dynamic interaction among institutional environment, government structure and the landlord port authority.

From the supply chain perspective, supply chain risk management (SCRM), as a major part of supply chain management, requires designing a dynamic supply chain network structure by considering both risks and uncertainties to manage the product flow throughout the system to predict, arrange and recover from disruptions (Baghalian, Rezapour, and Farahani, 2013). In addition, although it is acknowledged that both risks and uncertainties in SCRM are usually related to negative consequences in the most of literature (e.g. Wagner and Bode, 2006), the positive impacts posed by climate change can be considered. Therefore, how to take the advantage of these opportunities would be another consideration in dealing with the impacts in maritime transport and supply chain risk management. Although path dependence theory has been applied in studies of strategic port planning (Doom et al., 2013), historical conditions cannot serve as the only benchmark for decision-making in the future considering the tendency of more intensive, frequent, and unstable climate patterns. SCRM makes up this deficiency by emphasising the concept supply chain and quantifying the risks and potential uncertainties to design a dynamic supply chain network. Nevertheless, only SCRM may fail to answer the question of how to minimise the risks and uncertainties and maximise the benefits of stakeholders in the entire supply chains, which leads to the utilisation of stakeholder management.

Finally, stakeholder management reflects the demand of increasingly intensive competition and integration of multimodal port supply chains (Lam et al., 2013). However, it is still a challenge to coordinate all the stakeholders in adapting to the risks and uncertainties. As the impacts posed by climate change become more severe, it requires a comprehensive stakeholder management process where stakeholders emphasise the role, responsibility, and relevance of stakeholders, enhance the public participation in decision-making, and consider both risks and uncertainties in maritime transport and supply chains. The complexity of stakeholder management calls for the collaboration of multiple parties through joint projects and technological innovation. By initiating or improving a specific adaptation plan for climate change, it is expected to minimise the conflicts among stakeholders and maximise their common benefits in the future.

We believe that this chapter contributes to improve the planning process of maritime transport and supply chains, and address some of the deficiencies that might hinder the effective tackling of climate adaptation. It highlights various important structural principles of climate adaptation planning and existing loopholes that require paradigm shift solutions. The impacts posed by climate change to the world would likely become even more explicit in the foreseeable future, and continuous research germane to reduce uncertainty in decision-making dedicated to climate adaptation is necessary. We believe that the chapter offers a useful theoretical insight on how climate adaptation strategies and planning of ports can and should be developed.

References

Applegate, J. S. (2010). Embracing a precautionary approach to climate change. In D. M. Driesen (Ed.), *Economic thought and US climate change policy* (pp. 171–196). MIT Press.

Becker, A., Inoue, S., Fischer, M., & Schwegler, B. (2012). climate change impacts on international Seaports: Knowledge, perceptions, and planning efforts among Port administrators. *Climatic Change, 110*(1–2), 5–29.

Becker, A., Acciaro, M., Asariotis, R., Cabrera, E., Cretegny, L., Crist, P., Esteban, M., Mather, A., Messner, S., Naruse, S., Ng, A. K. Y., Rahmstorf, S., Savonis, M., Song, D. W., Stenek, V., & Velegrakis, A. F. (2013). A note on climate change adaptation for seaports: A challenge for global ports, a challenge for global society. *Climatic Change, 120*(4), 683–695.

Becker, A., Ng, A. K. Y., McEvoy, D., & Mullett, J. (2018). Implications of climate change for shipping: Ports and supply chains. *Wiley Interdisciplinary Reviews: Climate Change, 9*(2), wcc.508 (18).

Boxer, B. (1991). Societal contexts of ocean pollution science: Cross-national comparisons. *Global Environmental Change, 1*(2), 139–156.

Brooks, M. R., & Cullinane, K. (2007). Governance models defined. In *Devolution, port governance and port performance* (pp 417–448), Eds Idem. London: Elsevier.

Buitelaar, E., Lagendijk, A., & Jacobs, W. (2007). A theory of institutional change: Illustrated by Dutch City-provinces and Dutch land policy. *Environment and Planning A, 39*(2007), 891–908.

Christopher, M., & Peck, H. (2004). Building the resilient supply chain. *International Journal of Logistics Management, The, 15*(2), 1–14.

Dooms, M., Verbeke, A., & Haezendonck, E. (2013). Stakeholder management and path dependence in largescale transport infrastructure development: the port of Antwerp case (1960–2010). *Journal of Transport Geography, 27*, 14–25.

Fishman, R. (Ed.). (2000). *The American planning tradition: Culture and policy*. The Woodrow Wilson Center Press.

Fitzgerald, K. R. (2005). Big savings, but lots of risk. *Supply Chain Management Review, 9*, 9, 16–20.

Fuchs, M., & Scharmanski, A. (2009). Counteracting Path Dependencies: Rational Investment Decisions in the Globalising Commercial Property Market. *Environment and Planning A, 41*(11), 2724–2740.

Glassman, J. (2004). Transnational hegemony and US Labor Foreign Policy: Towards a Gramscian International Labour Geography. *Environment and Planning D, 22*(4), 573–593.

Gutmann, A., & Thompson, D. (2012). *The Spirit of compromise*. Princeton University Press.

Hall, P. A., & Taylor, R. C. R. (1998). (1998), political science and the three new institutionalisms. In K. Soltan, E. Soltan, & E. M. Uslaner (Eds.), *Institutions and social order* (pp. 14–44). University of Michigan Press.

Harvey, D. (2005). *A brief history of neoliberalism*. Oxford University Press.

Hodgson, G. (1993). *Economics and evolution: Bringing life Back in economics*. Polity.

Intergovernmental Panel on Climate Change (IPCC). (2012). *IPCC special report on managing the risks of extreme events and disasters to advance climate change adaptation*. IPCC.

Jiang, C., Zheng, S., Ng, A. K. Y., Ge, Y. E., & Fu, X. (2020). The climate change strategies of seaports: Mitigation vs. adaptation. *Transportation Research Part D: Transport and Environment, 89*, 102603.

Johnson, M. E. (2001). Learning from toys: lessons in managing supply chain risk from the toy industry. *California Management Review, 43*(3), 106–124.

Kay, A. (2005). A critique of the use of path dependency in policy studies. *Public Administration, 83*(3), 553–571.

Keohane, R. O., & Victor, D. G. (2010). *The regime complex for climate change, discussion paper, Harvard project on international climate agreements, Cambridge, MA* (January 2010) (no. 2010–33).

Khan, O., Christopher, M., & Burnes, B. (2008). The impact of product design on supply chain risk: a case study. *International Journal of Physical Distribution & Logistics Management, 38*(5), 412–432.

Lin, Y., Ng, A. K. Y., Zhang, A., Xu, Y., & He, Y. (2020). Climate change adaptation by ports: The attitude of Chinese port organizations. *Maritime Policy & Management, 47*(7), 873–884.

March, J. G., & Olsen, J. P. (1989). *Rediscovering institutions: The organizational basis of politics*. The Free Press.

Martin, R. (2000). (2000), institutional approaches in economic geography. In E. Sheppard & T. Barnes (Eds.), *A companion to economic geography* (pp. 77–94). Blackwell.

National Research Council (NRC). (2010). *Adapting to the impacts of climate change*. National Academies Press.

Ng, A. K. Y., & Liu, J. J. (2014). *Port-focal logistics and global supply chains*. Palgrave Macmillan.

Ng, A. K. Y., & Pallis, A. A. (2010). Port governance reforms in diversified institutional frameworks: Generic solutions, implementation asymmetries. *Environment and Planning A, 42*(9), 2147–2167.

Ng, A. K. Y., Becker, A., Cahoon, S., Chen, S. L., Earl, P., & Yang, Z. (Eds.) (2016). *Climate change and adaptation planning for ports*. Routledge. ISBN: 978-1-13-879790-1 (286 pages).

Ng, A. K. Y., Wang, T., Yang, Z., Li, K. X., & Jiang, C. (2018a). How is business adapting to climate change impacts appropriately? Some insight from the commercial port sector. *Journal of Business Ethics, 150*(4), 1029–1047.

Ng, A. K. Y., Zhang, H., Afenyo, M., Becker, A., Cahoon, S., Chen, S. L., Esteban, M., Ferrari, C., Lau, Y. Y., Lee, P. T. W., Monios, J., Tei, A., Yang, Z., & Acciaro, M. (2018b). Port decision-maker perceptions on the effectiveness of climate adaptation actions. *Coastal Management, 46*(3), 148–175.

Ng, A. K. Y., Monios, J., & Zhang, H. (2019). Climate adaptation management and institutional erosion: Insights from a major Canadian port. *Journal of Environmental Planning and Management, 62*(4), 586–610.

Norcliffe, G., Bassett, K., & Hoare, T. (1996). The emergence of postmodernism on the urban waterfront: geographical perspectives on changing relationships. *Journal of Transport Geography, 4*(2), 123–134.

North, D. C. (1990). *Institutions*. Cambridge University Press, Cambridge.

Notteboom, T., de Langen, P., & Jacobs, W. (2013). Institutional plasticity and path dependence in seaports: Interactions between institutions, port governance reforms and port authority routines. *Journal of Transport Geography, 27*(2013), 26–35.

Osthorst, W., & Manz, C. (2012). Types of cluster adaptation to climate change: Lessons from the port and logistics sector of Northwest Germany. *Maritime Policy & Management, 39*(2), 227–248.

Panahi, R., Ng, A. K. Y., & Pang, J. (2020). Climate change adaptation in the port industry: A complex of lingering research gaps and uncertainties. *Transport Policy, 95*, 10–29.

Paulsson, U. (2004). Supply chain risk management. *Supply Chain Risk: A Reader*, 79–96.

Pettigrew, A.M. (1992). The character and significance of strategy process research. *Strategic Management Journal, 13*(S2), 5–16.

Pierson, P. (2000). Increasing returns, path dependence, and the study of politics. *American Political Science Review, 94*(02), 251–267.

Poo, M. C.-P., Yang, Z., Dimitriu, D., & Qu, Z. (2018). Review on seaport and airport adaptation to climate change: A case on sea level rise and flooding. *Marine Technology Society Journal, 52*, 23–33.

Preston, B. L., Westaway, R. M., & Yuen, E. J. (2011). Climate adaptation planning in practice: An evaluation of adaptation plans from three developed Nations. *Mitigation and Adaptation Strategies for Global Change, 4*(2011), 407–438.

Sager, T. (2011). Neo-Liberal urban planning policies: A literature survey 1990-2010. *Progress in Planning, 76*(2011), 147–199.

Steinmo, S., Thelen, K., Longstreth, F. (Eds.), (1992). *Structuring politics: historical institutionalism and comparative analysis*. New York: Cambridge University Press.

Strambach, S. (2010). path dependency, path plasticity and the co-evolution of institutions and innovation: The German business software industry. In R. A. Boschma & R. Martin (Eds.), *Handbook for evolutionary economic geography* (pp. 406–431). Edward Elgar.

Tang, O., & Musa, S. N. (2011). Identifying risk issues and research advancements in supply chain risk management. *International Journal of Production Economics, 133*(1), 25–34.

UNCTAD. (2012). Ad hoc expert meeting on climate change impacts and adaptation: A challenge for global ports: Geneva, Palais des Nations, 29–30 September 2011: Main outcomes and summary of discussions, Geneva, Switzerland (September 2011).

UNECE. (2010). *Climate change impacts on international transport networks, working paper, working party on transport trends and economics*. UNECE.

United Nations. (2019). *Extreme weather hit 60 million people in 2018, no part of the world spared*. UN News: United Nations.

Van Driel, H., & Devos, G. (2007). Path dependence in ports: The persistence of cooperative forms. *Business History Review, 81*(04), 681–708.

Wagner, S.M., & Bode, C. (2006). An empirical investigation into supply chain vulnerability. *Journal of Purchasing and Supply Management, 12*(6), 301–312.

Weber, M. (1922). *Economy and Society*. (edited version by G. Roth and C. Wttich in 1978). University of California Press.

Wheeler, S. M. (2008). State and municipal climate change plans: The first generation. *Journal of the American Planning Association, 74*(4), 481–496.

Wheeler, S. M., Randolph, J., & London, J. B. (2009). Planning and climate change: An emerging research agenda. *Progress in Planning, 72*(2009), 210–222.

Xiao, Y., Fu, X., Ng, A. K. Y., & Zhang, A. (2015). Port investments on coastal and marine disasters prevention: Economic modeling and implications. *Transportation Research Part B: Methodological, 78*, 202–221.

Yang, Z., Ng, A. K. Y., Lee, P. T. W., Wang, T., Sanchez Rodrigues, V., Pettit, S., Harris, I., Zhang, D., & Lau, Y. Y. (2018). Risk and cost evaluation of port adaptation measures to climate change impacts. *Transportation Research Part D: Transport and Environment, 61*, 444–458.

Zsidisin, G. A. (2003). A grounded definition of supply risk. *Journal of Purchasing and Supply Management, 9*(5), 217–224.

Climate Risk Analysis: Uncertainty Modelling

1 Introduction

Climate change is at the forefront of research across disciplines as it is argued that it is an irreversible process which could pose catastrophic risks to human welfare (Keohane & Victor, 2010). Hence, the study of climate change is gradually moving away from purely mitigation towards a strategy of addressing mitigation and adaptation simultaneously (Ng et al., 2016). Maritime transport systems including both shipping and ports are highly vulnerable to climate risks such as flooding, hurricanes and sea level rise. Given the critical role that they play in the global economy and supply chains, analysing climate risk and their impacts on maritime transport systems poses a significant contemporary problem. However it is very challenging to use the traditional risk analysis methods (e.g. faulty tree analysis, failure mode, effect and criticality analysis) to address and evaluate such risks given the fact that compared to the classical risks (e.g. ship collisions) influencing maritime transport, climate change impact, as an emerging risk source, on maritime transport lacks of sufficient historical failure data to create a critical mass for effectively quantifying the frequency and consequences of different types of climate risks. As a result, advanced risk analysis models have to be developed to accommodate the high uncertainty introduced by the insufficient data.

Addressing risks of different features requires integration across hybrid uncertainty approaches and across research methodologies. Currently, we lack the critical understanding of which kinds of uncertainty models can most effectively harness science and technology for realizing climate risk analysis. The purpose of the chapter is to better understand the uncertainty in data posed by climate change on maritime systems, and how to effectively adapt to and manage such risks using the mainstream uncertainty theories (i.e. fuzzy possibilistic, Dempster-Shafer and Bayesian probabilistic) to develop new climate risk analysis models. This presentation will introduce a series of studies on climate risk analysis which enables the

Y.-y. Lau et al., *Maritime Transport and Supply Chain Resilience*,
https://doi.org/10.1007/978-3-032-07566-6_3

demonstration of the evolution of modelling climate risks facing maritime transport systems requiring data from a small to large scale. The work will address the significant methodological issues associated with risk analysis under high uncertainty particularly with reference to maritime transport. The findings will make significant contributions from two perspective. First, it strives to understand how maritime transport stakeholders can use advanced uncertainty modelling approaches to prioritise different types of climate risks facing maritime transport systems involving both operation and infrastructure. Second, it serves as the foundation to enable the development of a risk and cost evaluation methodology to generate the best practices and guidelines for providing appropriate long-term resilience and adaptation to climate change risks. The outcomes will be of value to port planners, policymakers and industrial practitioners, helping them to create and implement climate adaptation plans, strategies and practices.

The rest of the paper is structured as follows. The literature on climate risk parameters and three uncertainty modelling theories is presented in Sect. 2. The state-of-the-art climate risk analysis modelling approaches and their evolution in the past decade are described in Sect. 3. In this section, with regard to each discussed modelling approach, two new climate risk analysis and prioritisation approaches are introduced, including the description of climate risk data collection and process. In Sect. 4, discussion on alternative climate risk and resilience analysis approaches using objective data with lower level of uncertainty is carried out, along with future direction in terms of outlook of using artificial intelligence (AI) for climate risk analysis, before the conclusion in Sect. 5.

2 Literature Review

This section outlines the literature relating to climate risk parameters and three uncertainty modelling theories including fuzzy possibilistic, Dempster-Shafer and Bayesian probabilistic theories. The high uncertainty nature of the risk parameters in Sect. 2.1 justifies the demand of the literature on the three uncertainty theories in Sect. 2.2.

2.1 Climate Risk Parameters

A considerable number of methods and practices have been developed in recent years for identifying the vulnerabilities of transportation systems when facing the risks of climate change (Wang et al., 2020a). These studies were carried out from different perspectives, including the assessment of environmental impacts (i.e. Tonmoy & El-Zein, 2018; Matthews et al., 2017), economic analysis of climate risks and the associated adaptation costs (i.e. Qiao et al., 2015; Twerefou et al.,

2015), climate risk and resilience indicators (e.g. Poo et al., 2021a, b) and risk parameters focusing on uncertainty in data (e.g. Yang et al., 2018; Wang et al., 2020b).

In light of parameter driven climate risk analysis, many conventional risk assessment approaches (e.g. Quantitative Risk Assessment (QRA)) which have been widely used to carry out risk analysis across many sectors (Zhang et al., 2013), provide insights for identifying and defining the risk parameters used to define climate risks. Risk (*R*) is defined as a combination of the probability of occurrence (*L*) of an undesired event and the degree of its possible consequences (*C*) (Wang & Trbojevic, 2006), often referred as $R = L*C$. However, in history, this general risk definition has evolved to involve more risk parameters to better reflect the complexity of the described scenarios. For instance, when one intends to evaluate the risk of an airplane accident, *L* stands for the probability of an accident and *C* represent the consequence (i.e. all passengers onboard loss their lives), but the definition fails to describe the case where survivors exist in airplane accidents. Therefore, new parameters are introduced to enrich and empower the risk modelling approach for specific contexts. Among them, the probability of the failure being undetected (*P*) (also known as the probability of a consequence) is widely applied through a well-established risk analysis method, namely the Failure Mode and Effect Analysis (FMEA) approach, (Yang et al., 2008, 2009).

Based on the FMEA approach, scholars have proposed different risk parameters that can fit climate risk characteristics, including Yang et al., (2018), Wang et al., (2020b). They include (Yang et al., 2018):

- 'Timeframe (*T*)': How soon will a climate environmental hazard (e.g. sea level rise, flooding) occur. 'Likelihood (*L*)': how often will a climate environmental hazard occur given a specific *T*.
- 'Severity of Consequences (*C*)': how serious will the consequence caused by a climate environmental hazard be.

While *T* and *L* are relatively easier to understand and evaluate, *C* has more complicated features, affecting its effective measurement. First, the consequence (*C*) is determined by not only the damage ability of a climate environmental hazard but also the affected maritime transport system's resilience. It can be explained and reflected by the fact that the same climate environmental hazard will generate different impact onto different systems (e.g. ports in the same region). Therefore, a new parameter.

'Transport resilience' (Re) is introduced to complement the original *C* as follows (Wang et al., 2019, 2020b).

- 'Transport resilience' (Re): how resilient will a maritime transport system be when facing a climate environmental hazard and its impact.
  Accordingly, the original *C* is redefined as
- 'Severity of Consequences (*C*)': how serious will the damage ability of a climate environmental hazard be.
  In the meantime, in the serial studies, *C* is found to be difficult to directly evaluated as the damage ability is associated with three sub-parameters i.e.

environmental damage, economic loss and loss of lives. As a result, we have three new definitions.

- 'Damage to Infrastructure (INF)': how serious will the damage ability of a climate environmental hazard be, in terms of damage to infrastructure.
- 'Injuries and/or Loss of Lives (INJ)': how serious will the damage ability of a climate environmental hazard be, in terms of injuries and/or loss of lives.
- 'Damage to Environment (ENV)': how serious will the damage ability of a climate environmental hazard be, in terms of damage to environment.

As aforementioned, currently there are insufficient data to form a statistically recognised critical mass to quantify each of the above listed parameters and to calculate and estimate climate risk of any maritime transport system facing a specific climate environmental hazard. As a result, typical linguistic variables and their membership functions for the risk parameters are defined with reference to the work by Yang et al. (2009). Triangular/trapezoidal fuzzy membership functions developed on a [0–1] utility domain have been widely used to define different risk grades/descriptors. The typical membership functions for five linguistic variables are defined and characterised based on a balanced linear distribution in Tables 1, 2, 3 and 4 (Yang et al., 2009). It is possible to have some flexibility in the definition of membership functions to suit different risk scenarios. However, the changes of the defined membership functions require a very careful justification from domain experts. New definitions of memberships need to be verified through empirical tests before being applied in practice. The linguistics terms are suggested by domain experts through the Ad Hoc Expert Meetings organised by the United Nations Conference on Trade and Development (UNCTAD) (2012).

2.2 Three Uncertainty Modelling Theories

Uncertainty can be broadly categorised into three groups: fuzziness, incompleteness and randomness. Fuzziness refers to ill-defined, meaning that one cannot clearly define/describe an issue. For instance, the answer to the question of 'what is the temperature of a room' is normally given by cold, cool, warm or hot, instead of a precise numerical value. Incompleteness reveals the weak strength of evidence given background. For instance, the answer to the question of 'how sure do you think one can complete a task' is 80–90 percent, instead of 100%. Randomness represents the dynamic of an issue and its fast change nature. For instance, the answer to the question of 'how likely will a cyclist be involved in a road accident' is uncertain given the road, weather and traffic situations can change significantly from one time to another. With regard to the three uncertainty types, there are three mainstream uncertainty modelling theories, namely fuzzy logic, Dempster-Shafer (also known as evidential reasoning—ER) and Bayesian probabilistic theories.

Fuzziness is an aspect of vagueness and possibilistic uncertainty, which is indeed different from probabilistic one. Briefly, probability is a measure of the

Table 1 Timeframe

Grade	Linguistic terms	Description	Fuzzy memberships
1	Very short (VS)	Less than 1 year	(0, 0, 0.1, 0.3)
2	Short (S)	Approximately 5 years	(0.1, 0.3, 0.5)
3	Medium (M)	Approximately 10 years	(0.3, 0.5, 0.7)
4	Long (L)	Approximately 15 years	(0.5, 0.7, 0.9)
5	Very long (VL)	More than 20 years	(0.7, 0.9, 1, 1)

Source: Yang et al. (2018)

Table 2 Severity of consequence

Grade	Linguistic terms	Description	Fuzzy memberships
1	Catastrophic (CA)	Very severe economic loss and/or disruption on the facilities/systems/services requiring a very long period and very high cost of recovery	(0, 0, 0.1, 0.3)
2	Critical (CR)	Severe economic loss and/or disruption on the facilities/systems/services requiring a long period and long cost of recovery	(0.1, 0.3, 0.5)
3	Major (MA)	Significant economic loss and/or disruption on the facilities/systems/services requiring certain length of time and cost of recovery	(0.3, 0.5, 0.7)
4	Minor (MI)	Some economic loss and/or disruption on the facilities/systems/services requiring some time and cost of recovery	(0.5, 0.7, 0.9)
5	Negligible (NE)	A bit of disruption on the facilities/systems/services, and possibly with some economic loss, but with no real impacts on the continuance of services, nor does it require significant time and cost of recovery	(0.7, 0.9, 1, 1)

Source: Yang et al. (2018)

Table 3 Likelihood

Grade	Linguistic terms	Description	Fuzzy memberships
1	Very high (VH)	It is very highly likely that the stated effect will occur, with a probability around 90% of at least 1 such incident within the indicated timeframe	(0, 0, 0.1, 0.3)
2	High (H)	It is highly likely that the stated effect will occur, with a probability around 70% of at least 1 such incident within the indicated timeframe	(0.1, 0.3, 0.5)
3	Average (A)	It is likely that the stated effect will occur, with a probability around 50% of at least 1 such incident within the indicated timeframe	(0.3, 0.5, 0.7)
4	Low (L)	It is unlikely that the stated effect will occur, with a probability around 30% of at least 1 such incident within the indicated timeframe	(0.5, 0.7, 0.9)
5	Very low (VL)	It is very unlikely that the effects will occur, with a probability around 10% of at least 1 such incident within the indicated timeframe	(0.7, 0.9, 1, 1)

Source: Yang et al. (2018)

Table 4 Climate resilience

Grade	Linguistic terms	Capacity of transportation system	Time of recovery	Cost of recovery	Fuzzy memberships
1	Very weak (VW)	<20%	A year	>£10million	(0, 0, 0.1, 0.3)
2	Weak (W)	20–39%	A month	£1million-£ 9.9million	(0.1, 0.3, 0.5)
3	Average (A)	40–59%	A week	£100, 000-£999,999	(0.3, 0.5, 0.7)
4	Strong (S)	60–80%	A day	£10,000–£99,999	(0.5, 0.7, 0.9)
5	Very strong (VS)	>80%	12 hours	<£10,000	(0.7, 0.9, 1, 1)

Source: Wang et al. (2020b)

undecidability in the output of clearly defined and randomly occurring events, while fuzzy sets are concerned with the ambiguity inherent in the description of the event itself (Pan & McMichael, 1998). Informally, it can be said that the nature of possibility differs from a numerical variable and its values are not expressed using numbers but words in a natural or artificial language. Therefore, possibility theory, which emerged in the 1970s' (Zadeh, 1965), was developed quickly and became one of the most popular approaches in expert decision systems. Simultaneously, its uniqueness in dealing with vagueness, which differentiates with probability theory, leads to the inevitability of being widely applied in risk assessment. Jenson (2001) stated that except for probability theory, the most prominent approach to reasoning under uncertainty is possibility theory, which in certain contexts is called fuzzy logic. Fuzzy logic is an extension of classical Boolean logic from crisp sets to fuzzy sets. As a logic for reasoning, there is nothing fuzzy about fuzzy logic. Zadeh conceived the notion of fuzzy logic in 1965, the first new method of dealing with uncertainty since the development of probability (Zadeh, 1965). Since then, fuzzy logic has been a user-friendly profitable tool for analysing and controlling complex engineering systems and industrial processes (Deshpande, 1999) as well as knowledge-based expert decision systems. As a theoretical framework of expert decision making under uncertainty, fuzzy logic has been widely applied in the context of risk assessment (Wang et al., 1995; Pillay & Wang, 2003a).

The significance of fuzzy sets is that they facilitate gradual transition between states and consequently possess a natural capability to express and deal with observation and measurement uncertainties (Pillay & Wang, 2003b). Such a capability proves *FST* to be a useful tool in risk analyses as these analyses often require the use of subjective judgement and uncertain data. In principle, *FST* can restore integrity to the risk analyses by allowing uncertainty and not forcing precision where it is not possible. However, the theory can be difficult to apply without using linguistic variables as mediums. Since words in general are less precise than numbers, the concept of a linguistic variable serves the purpose of providing a means of approximate characterisation of phenomena, which are too complex or ill-defined to be amenable to description in conventional quantitative terms (Schmucker, 1984). The linguistic variables representing risk parameters are connected with fuzzy membership functions. Due to the advantage of simplicity, straight-line membership functions, especially triangular and trapezoidal membership functions have been commonly used to describe risks in safety assessment (Wang, 1997). After defining the membership functions of the risk related linguistic variables, the risk parameters can be assessed using subjective judgements. They can be expressed by either a complete linguistic term (a specific pre-defined fuzzy number), multiple linguistic terms with partial confidence or a new independent fuzzy number (i.e. approximately 20 percent). Furthermore, various fuzzy arithmetic calculations (i.e. the combination of Cartesian operation and decomposition operation, α-cuts and interval arithmetic, etc.) or rule-based definitions can be performed to obtain the fuzzy safety possibility estimations of the interested risk focuses. Finally, such fuzzy safety possibility estimations can be duffuzified to a crisp value, which can make more sense in risk ranking. Pure fuzzy logic has extremely limited applications (the only popularised application is the Sony Palmtop) and the main use of fuzzy logic is as an underlying logic system for fuzzy expert decision making systems (Pai, et al., 2003). In order to avoid the loss of useful information, a sign of attempting to synthesise fuzzy with ER and Bayesian network (BN) emerges as the direction of more scientifically based ways to assess risks.

The theory of evidence was first generated by Dempster (1967) and further developed by Shafer (1976). As such, the theory is often referred to as Dempster-Shafer theory of evidence or *D-S* theory. The *D-S* theory was originally used for information aggregation in expert systems as an approximate reasoning tool (Buchanan & Shortliffe, 1984; Lopez de Mantaras, 1990) and then used in decision making under uncertainty and risk in contrast to Bayes decision theory (Yager, 1992, 1995). *ER* is developed on the basis of the *D-S* theory. The use of *ER* as a risk analysis tool has been widely reported in the literature. Some typical studies making useful contributions towards the use of *ER* for representing and managing uncertainty in maritime risk include the works produced by Wang et al., 1995; Yang et al., 2009, 2014; Poo et al., 2021a, b.

An important achievement of applying *ER* to risk analysis is to incorporate both qualitative and quantitative risk parameters into one framework. When faced with a hybrid risk problem, the first thing to tackle is how to measure the qualitative risk parameter, followed by the question of how to combine it with quantitative risk parameters. An *ER* based approach with both qualitative and quantitative criteria under uncertainty was developed in the early 1990's (Yang & Singh, 1994; Yang & Sen, 1994). The major contribution of the approach lies in that it uses a distributed evaluation framework to overcome the inability of the *D-S* theory when conflicting evidence exists in problems. In the framework, assessments provided in terms of degrees of belief at lower-level parameters are aggregated through their weightings. Several applications of this approach in risk science can be addressed in the literature (Wang et al., 1995; Yang and Sen, 1996; Yang, 2001). The kernel of such an approach is an *ER* algorithm, which was generated by Yang and Singh, (1994), later updated by Yang and Sen (1994) and further modified by Yang (2001). The approach is continuously regenerated and its algorithm can be found and fully explained in Yang and Xu (2002) and the ER algorithm is presented below. The phrase '*ER* approach' throughout this thesis refers to the algorithm.

The set $S(E) = \{(H_n, \beta_n), n = 1, \ldots, N\}$ represents a parameter E which is assessed to grade $H_n$with degree of belief β_n, $n = 1, \ldots, N$. Let $m_{n,i}$ be a basic probability mass representing the degree to which the *ith* basic parameter e_i supports the hypothesis that the parameter y is assessed to the *nth* grade H_n. Therefore $m_{n,i}$ can be represented as follows (Yang & Xu, 2002; Yang, 2001):

$$m_{n,i} = \omega_i \beta_{n,i} \quad n = 1,2,\ldots,N; \quad i = 1,2,\ldots,L \tag{1}$$

$m_{H,i}$ is the remaining probability mass, that can be stated as:

$$\bar{m}_{H,i} = 1 - \sum_{n=1}^{N} m_{n,i} \qquad i = 1,2,\ldots\ldots,L \tag{2}$$

The remaining probability mass $m_{H,i}$ can be split into two parts $\bar{m}_{H,i}$ and $m_{H,i}$, which can be calculated by using the following formulas:

$$\bar{m}_{H,i} = 1 - \omega_i \qquad i = 1,2,\ldots,L \tag{3}$$

$$\tilde{m}_{H,i} = \omega_i \left(1 - \sum_{n=1}^{N} \beta_{n,i} \right) \qquad i = 1,2,\ldots,L \tag{4}$$

$\bar{m}_{H,i}$ is the first part of the remaining probability mass that is not yet assigned to individual grades due to the fact that parameter i (denoted by e_i) only plays one part in the assessment relative to its weight. $m_{H,i}$ is the second part of the remaining probability mass unassigned to individual grades, which is caused due to the incompleteness in the assessment $S(e_i)$.

To obtain the combined degrees of belief of all the basic parameter, $E_{I(i)}$ is firstly defined as the subset of the first i basic parameter as follows:

$$E_{I(i)} = \{e_1, e_2, \ldots, e_i\} \tag{5}$$

Let $m_{n,I(i)}$ be a probability mass defined as the degree to which all the i parameter in $E_{I(i)}$ support the hypothesis that E is assessed to the grade H_n and let $m_{H,I(i)}$ be the remaining probability mass unassigned to individual grades after all the basic parameter in $E_{I(i)}$ have been assessed. Eq. (3.6) and Eq. (3.7) are obviously correct when $i = 1$.

$$m_{n,I(1)} = m_{n,1}, n = 1,2,\ldots\ldots,N \tag{6}$$

$$m_{H,I(1)} = m_{H,1} \tag{7}$$

By using Eq. (3.6) and Eq. (3.7), Eq. (3.8) can be constructed for for $i = 1, 2, \ldots, L - 1$ to obtain the coefficients $m_{n, I(L)}$, $\bar{m}_{H,I(L)}$ and $m_{H,I(L)}$ (Yang & Xu, 2002):

$$K_{I(i+1)} = \left[1 - \sum_{t=1}^{N}\sum_{\substack{j=1 \\ j \neq t}}^{N} m_{t,I(i)} m_{j,i+1}\right]^{-1} \tag{8}$$

$K_{I(i+1)}$ is a normalizing factor.

$\{H_n\}$:

$$m_{n,I(i+1)} = K_{I(i+1)}\left[m_{n,I(i)} m_{n,i+1} + m_{H,I(i)} m_{n,i+1} + m_{n,I(i)} m_{H,i+1}\right] \quad n = 1,2,\ldots,N \tag{9}$$

$$\tilde{m}_{H,I(i+1)} = K_{I(i+1)}\left[\tilde{m}_{H,I(i)} + \bar{m}_{H,I(i)} \tilde{m}_{H,i+1} + \tilde{m}_{H\,I(i)} \bar{m}_{H,i+1}\right] \tag{10}$$

$$\bar{m}_{H,I(i+1)} = K_{I(i+1)} \bar{m}_{H\,I(i)} \bar{m}_{H,i+1} \tag{11}$$

$\{H\}$:

$$\bar{m}_{H,I(i)} = \tilde{m}_{H,I(i)} + \bar{m}_{H,I(i)} \qquad i = 1,2,\ldots,L-1 \tag{12}$$

At last, the combined degrees of belief of all the basic parameter for the assessment to parameter E are calculated by:

$$\{H_n\} : \beta_n = \frac{m_{n,I(L)}}{1 - \bar{m}_{H,I(L)}} \qquad n = 1,2,\ldots,N \tag{13}$$

$$\{H\} : \beta_H = \frac{\tilde{m}_{H,I(L)}}{1 - \bar{m}_{H,I(L)}} \tag{14}$$

Recently, *BNs* have led to many new applications of uncertainty modelling, in particular to very complex problems where a large of number of variables contribute to overall uncertainty. Such success of applying *BNs* stimulates and inspires researchers' interest of using *BNs* in system risk and reliability studies. Within risk analysis and safety decision management, the research related to *BNs* has been largely dominated in the scope of their practical applications. A *BN* (also called belief network, or probabilistic network) is a graphical presentation of probability combined with a mathematical inference calculation. It is used to represent dependencies between risk parameters. Each parameter represented as a node, is connected by directed links, represented as arrows or arcs, with conditional probability table (*CPT*) values assigned to the variables making up a *BN*. The nodes in a *BN* are called chance nodes. Chance nodes represent uncertain events or variables. The arrows are the directed links between nodes and this direction represents the conditional dependent relationship of these nodes.

The graphical representation makes *BNs* a flexible tool for constructing the models of causal impact between events, in particular when the causal impact has a random nature. Also, the specification of probabilities is focused on very small parts of the model (a variable and its parents). Having constructed the model, it can be used to compute effects of information as well as interventions from deterministic nodes. When the states of some nodes are fixed, the posterior probability distributions for the remaining nodes can be computed. Algorithms based on the Bayes' rule and Chain rule (Jensen, 2001) are developed for probability updating, and they perform very efficiently on a large variety of models. This makes *BNs* well suitable for risk forecasting and diagnosing.

A common criticism of the Bayesian approach is that it requires too much information in the form of prior probabilities, and that this information is often difficult or impossible to obtain in risk assessment. In practice, therefore, it is often necessary to rely on subjective probability estimates provided by expert judgements. It has been reported that linguistic expressions of probabilistic uncertainty may be more accurate than numerical values in estimating risk parameters (Wang et al., 1995) and that forcing the input of a crisp non-overlapping subjective probability is in many circumstances misleading and likely to lead to instability of the Bayesian system performance (Pan & McMichael, 1998). Consequently, an interesting circle (defined as *FBNs*) connecting *FST* with *BNs* may be formed. The use of fuzzy sets requires the support of *BNs* to deal with interactive dependence between risk parameters (randomness of uncertainty) and *BNs* rely on *FST* to cope with haziness of prior subjective probability (fuzziness of uncertainty).

3 Advanced Uncertainty Modelling for Climate Risk Analysis

This section outlines two methods of using advanced uncertainty modelling approaches to realise climate risk analysis and prioritisation in case of no objective data available to support the measurement of climate risk parameters. They are fuzzy rule-based ER approach in Sect. 3.1 and fuzzy rule-based BN approach in Sect. 3.2, respectively.

3.1 Fuzzy Rule Based Evidential Reasoning for Climate Risk Analysis

Fuzzy logic employing fuzzy *IF-THEN* rules, where the antecedent and conclusion parts contain linguistic variables of L, C and T (in Tables 1, 2 and 3), can model the qualitative aspects of human knowledge and reasoning process without employing precise quantitative analysis. To construct such systems in the context of climate risk analysis, the three risk parameters, *T*, *C* and *L* are considered as the antecedent attributes in *IF-THEN* rules. Risk level (*R*) is expressed as the conclusion attribute. A conventional *IF-THEN* rule in climate risk analysis can be expressed as follows.

$$\boldsymbol{Rule_k : IF\ t_k\ and\ c_k\ and\ l_k, THEN\ r_k}$$

where $\boldsymbol{t_k}$,$\boldsymbol{c_k}$and $\boldsymbol{l_k}$donate the linguistic variables of $\boldsymbol{T}$,$\boldsymbol{C}$*and* $\boldsymbol{L}$ used in the $\boldsymbol{k_{th}}$rule, $\boldsymbol{Rule_k}$, respectively; $\boldsymbol{r_k}$describes the consequence in $\boldsymbol{Rule_k}$ expressed by one single linguistic variable, describing $\boldsymbol{R}$ such as one of the set {Very High ($\boldsymbol{r_1}$), High ($\boldsymbol{r_2}$), Medium ($\boldsymbol{r_3}$), Low ($\boldsymbol{r_4}$), Very Low ($\boldsymbol{r_5}$)}.

Subjective degrees of belief (DoBs), $\beta_k^j\left(j = 1,2,\ldots,5\right)$ are assigned to the linguistic variables used to express the conclusion attribute *R* for modelling the incompleteness of expert judgements.

$$\boldsymbol{Rule_k : IF\ t_k\ and\ c_k\ and\ l_k, THEN}\ \left\{\left(\beta_k^1, r_1\right),\ldots,\left(\beta_k^5, r_5\right)\right\}\left(\sum_{j=1}^{5}\beta_k^j = 1\right)$$

A proportion method (Alyani et al., 2014) is used to rationalize the DoB distribution. Specifically, the DoB belonging to a particular grade in the *THEN* part is calculated by dividing the number of the risk parameters, which receive the same grade in the *IF* part in terms of fuzzy membership functions (e.g. VH and CA having the same fuzzy membership (0, 0, 0.1, 0.3)), by three. For instance, in a specific rule, the number of the risk parameters receiving the grade associated with a fuzzy membership (0, 0, 0.1, 0.3) in the IF part is two. The DoB belonging to the same grade (i.e. Very High) in the THEN part is therefore computed as 67% (2/3 × 100%). In a similar way, the FRB used in climate risk analysis containing 125 rules (5 × 5 × 5) with a rational DoB distribution is obtained and presented in Table 5.

After the *FRB* for climate risk analysis is constructed, it can be used to conduct risk inference using an *ER* technique. To start the inference process, observations available should be analysed to determine their relationship with each risk parameter in the antecedent in a numerical form. Four kinds of possible observations may be represented using membership functions to suit conditions under this study. They are either a single deterministic value with 100% certainty, a closed interval, a triangular distribution or a trapezoidal distribution (Wang, 1997). Having defined the fuzzy memberships of the four antecedent risk parameters, a matching function method (Liu et al., 2004) can be employed to convert the observations and determine the belief degrees to which the actual observations, which have been numerically described, match each linguistic variable in the antecedent. The matching function method chooses the *Max-Min* operation to show the similarity between the real input fuzzy set A^r and the corresponding fuzzy linguistic variables A_{iJ_i}, because it is a classical tool to set the matching degree between fuzzy sets (Zimmermann, 1991). Therefore, the matching degree between A^r and A_{iJ_i} can be defined as follows:

$$a_{iJi} = M\left(A^r, A_{iJi}\right) = \max_x \left[\min\left(\mu_{A^r}(x), \mu_{A_{iJi}}(x)\right)\right] \tag{15}$$

where x covers the domain of the input A^r. Each a_{iJ_i} represents the extent to which A^r belongs to the defined linguistic variables in the ith risk parameter in the antecedents. It is noteworthy that the fuzzy input may also be directly judged and expressed by experts using linguistic variables with belief structures.

Once the observations are converted and expressed by the pre-defined linguistic variables of the risk parameters with belief structures, it is necessary to introduce them into the rule-based inference system constructed in Table 5. An activating rule weight method is used to implement such an introduction. In other words, the distributions of different weights to all rules can be used to describe the relationship between the risk input transformed from observations and the rules in the system. In order to obtain an appropriate weight for the kth rule, the similarity degrees related to the kth rule are required to be synthesised in a logical way that can reflect the *AND* connective between their representing risk parameters. The *Product* operator as the logical tool has been recommended to model the *AND* connective in fuzzy rules (Liu et al., 2004), and thus, it is used to deal with the dependencies of the parameters T, L and C. Consequently, given that the three antecedent risk parameters have considered and assigned the same importance in this study, the weight of the kth rule can be calculated as follows:

$$\theta_k = \frac{\prod_{i=1}^{3} \alpha_{iJ_i}^{k}}{\sum_{l=1}^{125}\left(\prod_{i=1}^{3} \alpha_{iJ_i}^{l}\right)} \left(i = 1,\ 2,\ \text{or}\ 3;\ J_1, J_2\ \text{and}\ J_3 = 1,\ \ldots,\ \text{or}\ 5\right) \tag{16}$$

Note that the situations where some of $\alpha_{iJ_i}^{k}$ are equal to zero will significantly simplify the calculation through the ignorance of the rules including those linguistic variables with a zero similarity degree. A new rule base can be generated to include the relevant rules with weights more than zero. The significance of doing this is to reduce the size of the rule base investigated.

Having represented all the employed rules in the rule expression matrix, the *ER* approach (introduced in Sect. 2.2) and its attached calculation software IDS (Yang et al., 2001; Yang & Xu, 2002) can be used to combine the active rules (which replace the parameters given in the general ER algorithm description in Sect. 2.2.) and generate a final conclusion, which is a belief distribution on the climate risk expressions and will also give a panoramic view about the climate risk level for a climate environmental hazard. Similarly, when more climate risk parameters such as Re and the three sub-parameters of *C* are added, the new rule bases can be developed in a two-tire system to accommodate the additions. Basically, the current rule base in Table 5 will be extended from three antecedents to four antecedents (with Re added). Another rule base to simulate the IF-THEN relation between *C* and its three sub-parameters (i.e. INJ, INV and ENV) will also need to be developed using the same inference process. Obviously, it significantly increases computation cost every time when new parameters are added to climate risk analysis. New solutions to accelerate the inference process are wanted urgently. Section 3.2. addresses this demand by incorporating BN and using it to replace the ER inference.

3.2 *Fuzzy Rule-Based BN for Climate Risk Analysis*

Due to the high computation cost, the ER approach in Sect. 3.1 is replaced by BN in this section to introduce a new approach for climate risk analysis. Basically, the selected active rules in Sect. 3.1 will be synthesised by BN. The rule base with belief structures in Table 5 is firstly represented in the form of conditional probabilities and then further expressed in the form of conditional probability as follows.

Given $T1$, and $C1$, and $L3$, the probability of Rh ($h = 1,\ldots,5$) is $(1,0,0,0,0)$ or

$$p\left(Rh|,T1,C1,L1\right) = \left(1,0,0,0,0\right) \tag{17}$$

where '|' symbolizes conditional probability.

Using a *BN* technique, the *FRB* constructed in Table 5 can be modelled and converted into a four-node converging connection. It includes three parent nodes, N_T, N_C and N_L (Nodes T, C and L); and one child node N_R (Node R). Having transferred the rule base into a *BN* framework, the rule-based risk inference for the failure criticality analysis will be simplified as the calculation of the marginal probability of the node N_R. To marginalize R, the required conditional probability table of N_R, $p(S|L, C, P)$, can be obtained using (3.5), and the *FRB* shown in Table 4. It denotes a $5 \times 5 \times 5 \times 5$ table containing values $p(Rh|\,Ti, Cj, Lk)$ ($h, i, j, or\ k = 1,\ldots,5$).

The prior probabilities of N_T, $p(Ti)$ can be obtained through a questionnaire survey in which the question such as 'using the defined linguistic variables (i.e. VS, S, A, L and VL), how soon do you think a particular climate threat will have a negative

Table 5 FRB with belief structures for climate risk analysis

Rules	Antecedent attributes			Risk level (R)				
	Timeframe (T)	Severity of occurrence (C)	Likelihood (L)	Very high	High	Medium	Low	Very low
1	Very short (VH)	Catastrophic (CA)	Very high (VH)	100%	0	0	0	0
2	VH	CA	High (H)	67%	33%	0	0	0
3	VH	CA	Average (A)	67%	0	33%	0	0
…	…	…	…	…	…		…	…
123	Very long (VL)	Negligible (NE)	A	0	0	33%	0	67%
124	Very long (VL)	Negligible (NE)	Low (L)	0	0	0	33%	67%
125	Very long (VL)	Negligible (NE)	Very low (VL)	0	0	0	0	100%

Source: Yang et al. (2018)

impact on port?' will be developed to collect the domain experts' risk perceptions. The results from multiple experts across ports and regions will be classified based on investigated ports and regions, purified to avoid the input of obvious subjective bias and finally aggregated using an ER calculation. In a similar way, the prior probabilities of N_C, and N_L can be computed as *p(Cj)*, and *p(Lk)*, respectively. Having analysed all the prior probabilities of the four nodes, the marginal probability of N_R can be calculated as (Jensen, 2001):

$$p(Rh) = \sum_{i=1}^{5} \sum_{j=1}^{5} \sum_{k=1}^{5} p(Rh|,Ti,Cj,Lk)\, p(Ti)\, p(Cj)\, p(Lk) \; (h = 1,\ldots,5) \tag{18}$$

To prioritize the climate risks, *Rh* ($h = 1, \ldots, 5$) requires the assignment of appropriate utility values U_{Rh}. The utility values can be defined on the basis of a linear distribution as *Rh* ($h = 1, \ldots, 5$) = $\{0, 0.25, 0.5, 0.75, 1\}$.

Then a new risk priority/ranking index can be developed as:

$$RI = \sum_{h=1}^{5} p(Rh)\, U_{Rh} \tag{19}$$

where the larger the value of *RI* is the lower the risk level of potential climate environmental hazards. The calculation process of *RI* can be simplified by using a *Hugin* software package (Andersen et al., 1990).

After the introduction of the above fuzzy rule-based BN methodology, it can be applied to provide an effective solution to more complicated climate risk analysis involving new risk parameters such as the aforementioned Re, INJ, INV and ENV, and use Hugin software to realise the automation of the analysis and ranking. While Figs. 1 and 2 provides the structure of the developed FRB and BN model respectively, the technical details of the model development is documented in Wang et al.

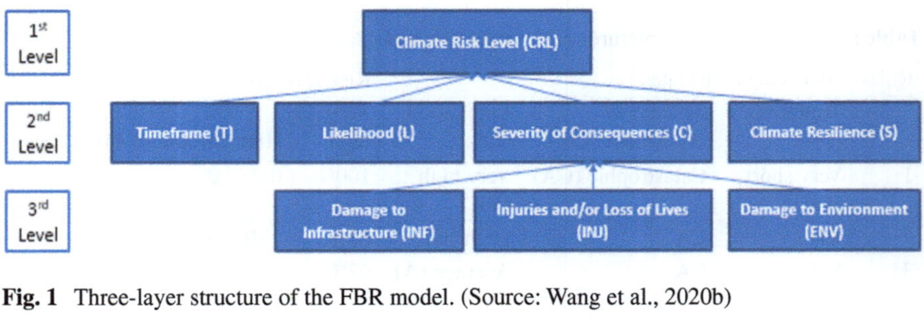

Fig. 1 Three-layer structure of the FBR model. (Source: Wang et al., 2020b)

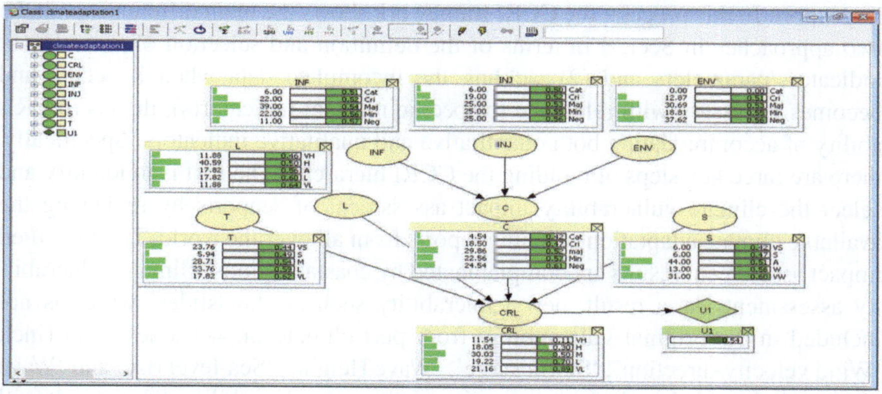

Fig. 2 Climate risk analysis of a environmental hazard using Hugin. (Source: Wang et al., 2020b)

4 Objective Data-Driven Climate Risk Analysis Approaches and a New Thought on Future Direction

The high uncertainty in data for climate risk analysis mainly comes from insufficient data for measuring the defined risk parameters. The two approaches in Sect. 3 are criticised due to their high dependency on subjective data, which could arguably introduce subjective bias for the findings, i.e. the ranking of identified climate environmental hazards influencing maritime transport systems. To address this shortcoming, new objective data driven climate risk analysis approaches which involve lower level uncertainty in terms of input data subjectivity are proposed and introduced in this section. Within this context, the representative works are climate change risk indicators (CCRI) and climate resilience indicators (CRI) work by Poo et al. (2021a, b), respectively. Finally, this section describes a new through on data driven climate risk analysis using AI methods.

Maritime transport systems such as seaports are exposed to the vulnerability of climate impacts, mainly because of their locations, including low-lying areas, coastal zones and deltas. The section describes the development of a CCRI framework for climate risk assessment of seaports, enabling research-informed policy-making on such a demanding topic. Climate risk analysis aids rational adaptation planning. Many climate assessments have been done for measuring climate risks (e.g. Sect. 3), and various climate adaptation measures have been proposed for reducing climate risks. However, few of them used objective data for climate risk evaluations in seaports and fewer on the provisions of CCRIs for comparing climate risks of different locations and timeframes to guide rational policy making. Furthermore, climate change is a dynamic issue, requiring big objective data to support the analysis (e.g. monthly climate data on CCRIs) of climate threats and vulnerabilities.

To do so, ER is employed to evaluate the climate risks in seaports by (1) synthesising the CCRIs from lower level to the top level to obtain the climate risk of a target port. In this process, the CCRI framework shows the main difference with the two approaches in Sect. 3 in terms of the definition and selection of the relevant indicators/parameters and (2) tackling the incomplete data when it occurs and becomes necessary with reference to specific relevant indicator(s), thanks to ER's ability of accommodating both quantitative and qualitative indicators. Specifically, there are three key steps of building the CCRI hierarchy. The first is to identify and select the climate vulnerability impact assessment of seaports by reviewing the available climate adaptation planning reports from all over the world. The identified impact assessment results are complemented by coastal regions' climate vulnerability assessment. As a result, new vulnerability such as 'landslides' which is not included in the original vulnerability from port climate impact assessment (incl. 'Wind velocity/direction', 'Storm surge', 'Wave Height', 'Sea-level rise' and 'Wave direction') is added to form the new list of climate vulnerability. Such result will then be used to guide to identify the climate change risk indicators such as 'extreme precipitation'. The second step is to collect all seaport risk items from Department for Environment, Food & Rural Affairs (DEFRA) who invited nine UK seaport professional bodies. They had submitted climate change adaptation reports about seaport risks under Climate Change Act 2008. As a result, 334 risk items have been identified and addressed with different formats and scales. Even though the risk levels of each item cannot be directly compared, some insights can still be observed by statistical analyses and by visualising the climate vulnerabilities in this country. Three sets of categories have been set up by Poo et al., (2021a) manually, including climate threat types, seasons and operation sectors. To define them on a standardised plate, different EWEs are reclassified with reference to the categories by the IPCC working group II in the Fifth Assessment Report of 2014, including 'Extreme precipitation', 'Heat wave/High temperature', 'Cold wave/Increased snow events', 'Sea-level rise (SLR)/Storm surge' and 'Storminess' (IPCC, 2014). Finally, the obtained CCRIs are matched with the Met Office climate data. To maintain the objectivity of the data (for the purpose of addressing the data subjectivity of the two approaches in Sect. 3), only CCRIs which have the objective climate data from the Met office available are selected to form the CCRI hierarchy as shown in Fig. 3.

After the development of the CCRI hierarchy, each indicator has to be graded. In this process, As monthly average climate data represent the possibilities of EWEs, percentile values of climate data are commonly used in assessing climate vulnerability (Monahan & Fisichelli, 2014; Peterson et al., 2002). Percentile values of different CCRIs, shown in Table 8, are chosen and the dataset for the CCRI framework is set up by the reference from Task Team On Definitions Of Extreme Weather And Climate Events (2016). Finally, one port's climate data is collected and converted into an expression by the defined grades with DoB using a distance approach (Poo et al., 2021a). The ER algorithm (i.e. Eqs. (3.1–3.14) are then applied to synthesise the port's performance value against each bottom-level indicators to calculate and obtain its CCRI. The findings reveal the quantitative measures of climate change risks in different locations and in different months. Furthermore, the risk levels of seaports in the future are assessed for observing the changes and informing policy making. The main contributions of this study include the visualisation of the comprehensive climate risk levels and *provision* of a new climate risk analysis

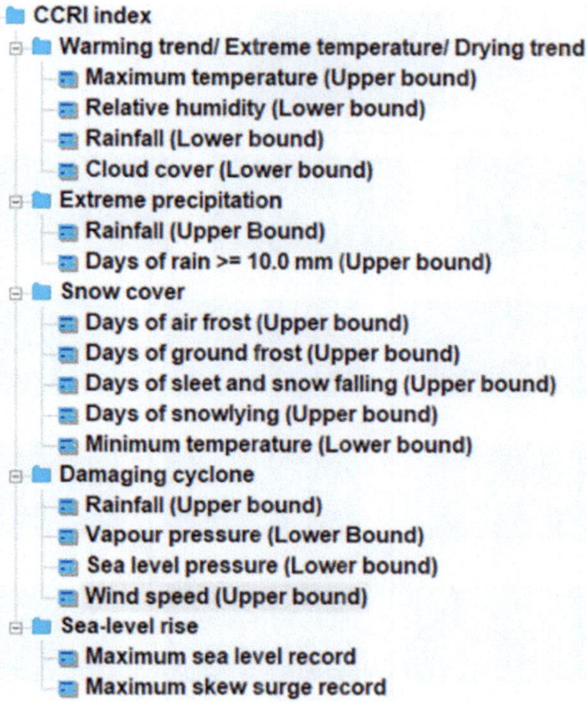

Fig. 3 CCRI hierarchy. (Source: Poo et al., 2021a)

framework through the comparison of climate change risks with respect to different locations and timeframes. Suitable climate adaptation measures can be chosen to implement, and seaports can cooperate on climate resilience issues (e.g. seaport network service and pre-disaster relief logistics).

Climate risk analysis research development comes across a pattern of first configuring climate change risk (with a focus on external environmental hazards) and then taking into account the resilience of transport infrastructures (concerning the impacted target resilience ability to reflect the different impacts across different infrastructures when facing the same external environmental hazard). The subjective data-oriented approaches in Sect. 3 follow it and the objective data driven solutions are not exceptional. Following the above CCRI framework, Poo et al., (2021b) introduced a new CRI framework. While the main principle is identical to the aforementioned CCRI steps, the main difference lies the development of CRI hierarchy as shown in Fig. 4. It is noteworthy that although the CRI in Fig. 4 was designed for airport climate resilience analysis, it can be adapted to seaports without the need of making major methodological changes as airports and seaports share commonalities in terms of climate risk analysis.

Inspired by the increasing use of AI solution in classical risk (e.g. ship collisions) analysis of maritime transport systems, we propose a new thought for facilitating climate risk analysis using big data. It enables the movement of climate risk analysis approaches with high uncertainty in data (Sect. 3), through the ones with low

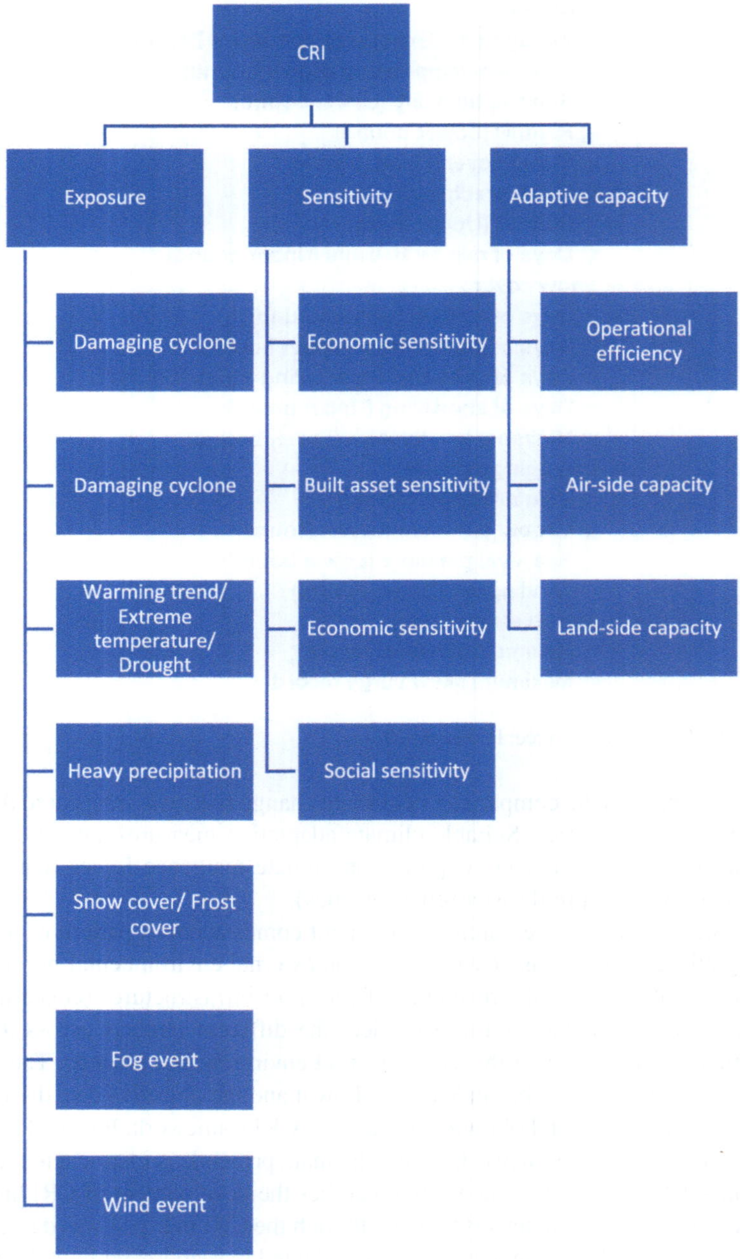

Fig. 3.4 The CRI framework. (Source: Poo et al., 2021b)

uncertainty in data (Sect. 4), to the ones with big data for a future new solution. The idea is to use text mining approaches to screen all climate disasters to find out the hidden parameters/factors influencing disaster occurrence which have yet been identified in the existing literature. The result will then need to match with the investigated maritime transport infrastructures and/or operations to formulate the association between the factors and climate risk impact consequence. The gathering of the association from all the reports will contribute to a new database which can then be used to train a machine learning or deep learning mechanism for climate risk prediction and/or diagnosis in future.

5 Conclusion

This chapter introduces three main uncertainty modelling theories and demonstrates their applications to realise climate risk analysis and quantification under different levels of uncertainty in data. The approaches can aid decision making to prioritise the climate risk facing their concerned vulnerability maritime transport systems first (and other transport systems later from a supply chain perspective). The results will make significant contributions to rationalise the development of risk-based climate adaptation planning in which all measures can be assured for their cost effectiveness. For this reason, future work could continue to develop new climate adaptation planning for maritime transport systems based on the findings from this chapter, to improve the systems' overall resilience. It will also make new contribution to providing quantitative climate risk values to enable the systems risk evaluations from all different perspectives which will help justify the high importance of addressing climate risks over the other classical ones. It can help provide scientific evidence to change decision makers' perception that climate risk is remote and uncertain while the classical ones (e.g. ship collisions) shall be given priority in terms of safety resource allocation. Within this context, the chapter push the climate risk analysis methodology to a new high level, where climate risk can be directly compared with other hazards influencing maritime transport safety for attraction of sufficient attention, politically and financially.

Acknowledgements This project has received funding from the European Research Council (ERC) under the European Union's Horizon 2020 research and innovation programme (Grant Agreement No. 864724).

References

Alyami, H., Lee, P. T. W., Yang, Z., Riahi, M., Bonsall, S., Wang, J. (2014). An advanced risk analysis approach for container port safety evaluation. *Maritime Policy & Management, 41* (7), 634–650.

Andersen, S.K., Olesen, K.G., Jensen, F.V. and Jensen, F. (1990), "Hugin – A Shell for Building Belief Universes for Expert Systems", Reading in Uncertainty, Shafer, G. and Pearl, J. (eds.), Morgan Kaufmann, pp. 332–337.

Buchanan, B. G., & Shortliffe, E. H. (1984). *Rule-based expert systems*. Addison-Wesley.

Dempster, A. P. (1967). Upper and lower probabilities induced by a multivalued mapping. *Annals of Mathematical Statistics, 38*, 325–339.

Deshpande, A. W. (1999). Application of fuzzy set theory to environmental engineering systems. In: *BISC seminar, Oct. 7*, National Environmental Engineering Research Institute (NEERI), India.
IPCC. (2014). In C. B. Field, V. R. Barros, D. J. Dokken, K. J. Mach, M. D. Mastrandrea, T. E. Bilir, M. Chatterjee, K. L. Ebi, Y. O. Estrada, R. C. Genova, B. Girma, E. S. Kissel, A. N. Levy, S. MacCracken, P. R. Mastrandrea, & L. L. White (Eds.), *Impacts, adaptation, and vulnerability. Part A: Global and sectoral aspects. Summary for policymakers*. Cambridge University Press.
Jensen, F. V. (2001). *Bayesian network and decision graphs*. Springer.
Keohane, R. O., & Victor, D. G. (2010). *The regime complex for climate change*. Discussion paper 2010–33. The Harvard Project on International Climate Agreements, Cambridge, MA, January 2010.
Liu, J., Yang, J, B., Wang, J., SII, H. S., Wang, Y. M., (2004). Fuzzy rule-based evidential reasoning approach for safety analysis. *International Journal of General Systems, 33* (2–3), 183–204.
Lopez, D. M. R. (1990). *Approximate reasoning models*. Ellis Harwood Ltd.
Matthews, L., Andrey, J., & Picketts, I. (2017). Planning for winter road maintenance in the context of climate change. *Weather, Climate, and Society, 9*(3), 521–532.
Monahan, W. B., & Fisichelli, N. A. (2014). Climate exposure of US national parks in a new era of change. *PLoS One, 9*, e101302.
Ng, A. K. Y., Becker, A., Cahoon, S., Chen, S. L., Earl, P., & Yang, Z. (Eds.). (2016). *Climate change and adaptation planning for ports*. Routledge.
Pai, R. R., Kallepalli, V. R., Caudill, R. J., & Zhou, M. C. (2003). Methods towards supply chain risk analysis. *System Security and Assurance, 5*, 4560–4565.
Pan, H. P., & McMichael, D. (1998). Fuzzy causal probabilistic networks—A new ideal and practical inference engine. In: *Proceeding of the 1st international conference on multisource-multisensor data FUSION, FUSION 98*, July, Las Vegas, NV, USA.
Peterson, T. C., Taylor, M. A., Demeritte, R., Duncombe, D. L., Burton, S., Thompson, F., Porter, A., Mercedes, M., Villegas, E., & Semexant Fils, R. (2002). Recent changes in climate extremes in the Caribbean region. *Journal of Geophysical Research: Atmospheres, 107*, ACL 16-1–ACL 16-9.
Pillay, A., & Wang, J. (2003a). *Technology and safety of marine systems*. Elsevier.
Pillay, A., & Wang, J. (2003b). A risk ranking approach incorporating fuzzy set theory and grey theory. *Engineering Reliability & System Safety, 79*(1), 61–67.
Poo, M. C. P., Yang, Z., Dimitriu, D., Qu, Z., Jin, Z., & Feng, X. (2021a). Climate change risk indicators (CCRI) for seaports in the United Kingdom Ocean & Coastal Management 205, 105580.
Poo, M. C. P., Yang, Z., Dimitriu, D., Qu, Z., (2021b). An advanced climate resilience indicator framework for airports: A UK case study. Transportation Research Part D: Transport and Environment 101, 103099.
Qiao, Y., Dawson, A. R., Parry, T., & Flintsch, G. W. (2015). Evaluating the effects of climate change on road maintenance intervention strategies and life-cycle costs. *Transportation Research Part D: Transport and Environment, 41*, 492–503.
Schmucker, K. J. (1984). *Fuzzy sets, natural language computations and risk analysis*. Computer Science Press.
Shafer, G. (1976). *A mathematical theory of evidence*. Princeton University Press.
Tonmoy, F. N., & El-Zein, A. (2018). Vulnerability to sea level rise: a novel local-scale indicator-based assessment methodology and application to eight beaches in Shoalhaven, Australia. *Ecological Indicators, 85*, 295–307.
Twerefou, D. K., Chinowsky, P., Adjei-Mantey, K., & Strzepek, N. L. (2015). The economic impact of climate change on road infrastructure in Ghana. *Sustainability, 7*(9), 11949–11966.
UNCTAD. (2012). *Ad hoc expert meeting on climate change impacts and adaptation: A challenge for global ports: Geneva, Palais Des Nations, 29–30 September 2011: Main outcomes and summary of discussions*. UNCTAD.
Wang, L. X. (1997). *A course in fuzzy systems and control*. Prentice-Hall.
Wang, J., & Trbojevic, V. M. (2006). *Design for Safety of large marine and offshore engineering products*. Institute of Marine Engineering, Science and Technology (IMarEST).

Wang, J., Yang, J. B., & Sen, P. (1995). Safety analysis and synthesis using fuzzy set modelling and evidential reasoning. *Reliability Engineering & System Safety, 47*(3), 103–118.

Wang, T., Qu, Z., Yang, Z., Nichol, T., Dimitriu, D., Clarke, G., & Bowden, D. (2020a). How can the UK road system be adapted to the impacts posed by climate change? By creating a climate adaptation framework. *Transportation Research Part D: Transport and Environment, 77*, 403–424.

Wang, T., Qu, Z., Yang, Z., Nichol, T., Dimitriu, D., Clarke, G., Bowden, D., & Lee, P. T. (2020b). Impact analysis of climate change on rail systems for adaptation planning: A UK case. *Transportation Research Part D: Transport and Environment, 83*, 102324.

Yager, R. R. (1992). On the specificity of a possibility distribution. *Fuzzy Sets and Systems, 50*, 279–292.

Yager, R. R. (1995). An approach to ordinal decision making. *International Journal of Approximate Reasoning, 12*, 237–261.

Yang, J. B. (2001). Rule and utility based evidential reasoning approach for multiattribute decision analysis under uncertainties. *European Journal of Operational Research, 131*(1), 31–61.

Yang, J. B., & Sen, P. (1994). A general multi-level evaluation process for hybrid MADM with uncertainty. *IEEE Transaction on Systems, Man and Cybernetics, 34*(10), 1458–1473.

Yang, J. B., & Singh, M. (1994). An evidential reasoning approach for multiple attribute decision making with uncertainty. *IEEE Transaction on Systems, Man and Cybernetics, 24*(1), 1–18.

Yang, J. B., & Xu, D. L. (2002). On the evidential reasoning algorithm for multiple attribute decision analysis under uncertainty. *IEEE Transactions on Systems, Man and Cybernetics—Part A: Systems and Humans, 32*(3), 289–304.

Yang, Z., Bonsall, S., & Wang, J. (2008). Fuzzy rule-based Bayesian reasoning approach for prioritization of failures in FMEA. *IEEE Transactions on Reliability, 57*(3), 517–528.

Yang, Z. L., Bonsall, S., & Wang, J. (2009). Use of fuzzy evidential reasoning in maritime security assessment. *Risk Analysis, 29*(1), 95–120.

Yang, Z., Ng, A. K. Y., & Wang, J. (2014). Incorporating quantitative risk analysis in port facility security assessment. *Transportation Research Part A: Policy and Practice, 59*, 72–90.

Yang, Z., Ng, A. K. Y., Lee, P. T. W., Wang, T., Qu, Z., Rodrigues, V. S., Pettit, S., & Harris, I. (2018). Risk and cost evaluation of port adaptation measures to climate change impacts. *Transportation Research Part D: Transport and Environment, 61*, 444–458.

Yang et al. (2001). Rule and utility based evidential reasoning approach for multiattribute decision analysis under uncertainties JB Yang *European Journal of Operational Research, 131* (1), 31–61.

Yang, J. B., & Sen, P. (1996). A general multi-level evaluation process for hybrid MADM with uncertainty. IEEE Transactions on Systems, Man, and Cybernetics 24 (10), 1458–1473.

Zadeh, L. A. (1965). Fuzzy Sets. *Information and Control, 8*(3), 338–353.

Zhang, D., Yan, X. P., Yang, Z. L., Wall, A., & Wang, J. (2013). Incorporation of formal safety assessment and Bayesian network in navigational risk estimation of Yangtze River. *Reliability Engineering & System Safety, 118*, 93–105.

Zimmermann 1991. Zimmermann, H.J. (1991), Fuzzy Set Theory and its Application, Norwell, MA, Kluwer. Andersen et al (1990). Andersen, S.K., Olesen, K.G., Jensen. F.V. and Jensen, F. (1990), "Hugin – A Shell for Building Belief Universes for Expert Systems", Reading in Uncertainty, Shafer, G. and Pearl, J. (eds.), Morgan Kaufmann, pp. 332–337.

Port Centrality and Resilience Analysis for Global Shipping Network in the Context of Typhoons

1 Introduction

The shipping industry is the artery of the global economy, undertaking important tasks in international trade and the global supply chain. According to the World Class Marine Port Cluster Composite Index Report (2024) (Comnews, 2024), the Central and Southern Yangtze River Delta Port Cluster, Shandong Port Cluster and Pearl River Delta Port Cluster in China are on par with the international port clusters of Japan's Tokyo Bay Port Cluster and the New York-New Jersey Port Cluster, are together ranked among the top of the world-class marine port clusters. These port clusters not only dominate in terms of physical facilities but have also made remarkable achievements in regional cooperation and the construction of integrated service systems.

By the end of 2023, the operational capacity of China's ports had been significantly improved, as evidenced by the total number of berths owned by production terminals reaching 22,023, of which 2878 berths were capable of accommodating ships of more than 10,000 tons. During the year, the cargo throughput of Chinese ports reached an astonishing 16.973 billion tons, and the container throughput also reached 310 million TEUs, ranking first in the world for many years. Of the world's 50 largest ports in terms of cargo throughput, Chinese ports occupy 28 seats, and the top six are all Chinese ports, which undoubtedly further validates China's leading position in the global shipping industry (Ministry of Transport of the People's Republic of China,2024).

China's shipping industry has significant competitiveness and influence globally, but it also faces the challenge of extreme weather events. According to data from the Tropical Cyclone Data Center of the China Meteorological Administration, 661 typhoons made landfall in China, with an average of eight to nine affecting the country each year; the number of landfalls totaled 893, with an average annual number of landfalls reaching 11.91 from years 1949 to 2023. Given the significance of

Y.-y. Lau et al., *Maritime Transport and Supply Chain Resilience*,
https://doi.org/10.1007/978-3-032-07566-6_4

the impact of typhoons on China's coastal ports, researchers have been investigating how to respond to these extreme weather events to minimise their impact on the shipping network.

In order to strengthen the sustainable development of the shipping industry, the assessment of the importance and resilience of the port in its shipping network is particularly significant. In this chapter, the authors first select five typical Chinese ports from each port cluster of China and then conduct an in-depth analysis on their centrality within their shipping network, together with the assessment of the resilience of the top two critical ports considering the influence of unprecedented risks, namely typhoons, of their global shipping networks.

2 Literature Review

2.1 Impacts of Typhoon Disasters on Ports

In terms of port typhoon risk assessment, the impact of typhoons on ports has been analysed by constructing assessment systems, collecting historical data and simulating typhoon scenarios using different models. At the same time, some scholars have researched reducing typhoons' impact on ports, exploring how to minimise the impacts and losses on ports.

Yang and Ge (2020) created a framework consisting of ten indicators, collected expert questionnaire data, evaluated the vulnerability of Kaohsiung Port to typhoons, developed specialised vulnerability response measures, and provided important reference significance. Mou et al. (2024) analysed AIS trajectory data and typhoon data to assess typhoon disruptions in China's key ports and applied entropy weighting and grey relation analysis to study the effects of removing MSR's critical Chinese ports on container shipping. Hong and Ji (2019) gathered meteorological data on typhoons in the Nansha port and analysed disaster levels across various recurrence periods. The study integrated local risk knowledge into GIS software, offering a reference for similar work. Huang et al. (2024) analysed 776 million AIS signals to assess typhoon effects on container shipping networks, revealing changes in network structure, port importance, vessel traffic and route distance fluctuations. Yuan et al. (2023) simulated typhoon impacts on port systems using container supply capacity as a metric, integrating operational data to conduct risk analyses for individual ports and clusters. Hsieh (2014) developed a risk assessment framework that integrates physical, economic and social port factors, assessing facility vulnerability, operational complexity and external shock sensitivity. Gao et al. (2020) employed an energy-value method to assess regional typhoon risks by comparing energy flows to an everyday solar energy basis. They built an indicator system for resilience, intensity and sensitivity using 2017 typhoon data, appls it to evaluate typhoon risks in Zhuhai, Guangdong, China. Wang et al. (2022) used geographic information systems (GIS) and related risk assessment models to investigate regional typhoon disaster risks in space and time. He also discussed the importance of typhoon disaster risk assessment and proposed corresponding risk management and mitigation strategies. Liu et al. (2023) introduced a 3D information diffusion-based typhoon risk assessment method, integrating historical typhoon data, GIS and disaster impacts into a multi-dimensional model to enhance risk comprehension and prediction for improved disaster management and mitigation. Wan et al. (2024) introduced a fuzzy evidence reasoning model for evaluating the typhoon resistance of ports during extreme weather conditions, which drew on relevant literature and staff surveys. Wang et al. (2024) took Zhoushan Port in Ningbo, China, as an example to assess the typhoon risk level of container ports, predict their economic losses and evaluate corresponding adaptation measures.

In summary, typhoon impact studies on ports address two main issues: evaluating port vulnerability and risk and finding mitigation strategies. Current research often overlooks inter-port interactions and cascading effects across shipping networks. To address this gap, this chapter introduces five port centrality indicators to assess ports' importance in the maritime network, enhancing the comprehensive evaluation of port resilience to extreme weather.

2.2 *Port Resilience*

Port resilience is the capacity to endure, recover and adapt from shocks like typhoons, earthquakes and floods. It encompasses infrastructure strength, operational flexibility and economic resilience. As a key port sustainability and safety measure, resilience impacts disaster response efficiency and reconstruction speed.

Shafiezadeh and Burden (2014) developed a scenario-based framework for assessing infrastructure resilience and evaluating port infrastructure based on disaster intensity, providing a systematic approach. Lin et al. (2022) evaluated the impact of the Epidemic Prevention Alliance Strategy (EPAS), Shared Berth Strategy (SBS) and Hybrid Strategy on port congestion and created a dual-port system dynamics model to measure port resilience. Wang et al. (2023) introduced a cyclic four-stage method to analyse port elasticity, using Bayesian models to identify interruption factors and evaluate the elasticity of Yangshan Port. Kim et al. (2021) developed a port resilience measurement framework focusing on Korean ports' resilience to disruptions. The model encompasses infrastructure, operations, economics and socio-environmental resilience. Using system dynamics, it simulates port responses and recoveries in various scenarios. Panahi et al. (2022) created a Bayesian network model to assess port resilience amidst COVID-19, identifying key operational factors and evaluating their effects on resilience based on expert insights and statistical data. Tian et al. (2024) presented an integrated framework combining influence diagrams and physics-based modelling to assess tropical cyclone impacts on seaports. Six different historical typhoon scenarios are used to compare the actual operation state and the produced results. Cho and Park (2017) developed a system dynamics model for assessing South Korean port infrastructure resilience to earthquakes and tsunamis. The model evaluates damage, recovery and socioeconomic changes, simulating disaster scenarios to analyse recovery processes and economic impacts, aiding in resilience comparison and policy formulation.

In current research on port resilience, there is often a lack of evaluation systems based on specific indicators, which hinders a comprehensive understanding of port resilience and overall performance evaluation in emergency situations. The existing system is incomplete and neglects key aspects such as infrastructure and resource allocation, which hinders port management personnel from comprehensively assessing resilience and affects planning and decision-making efficiency.

In summary, the lack of a comprehensive evaluation framework for port resilience research under typhoons hinders managers' comprehensive decision-making. Current research often focuses only on a single port and neglects the broader maritime network. This chapter first constructs a shipping network based on the five major ports in China and evaluates their importance. Then, it divides port resilience assessment into "absorption phase," "adaptation phase," and "recovery phase" to deepen understanding. Finally, the top two ports are selected from the evaluation results of the five major ports in China, and their toughness is studied by AHP-entropy weight method and TOPSIS method, considering the subjective and objective weights.

3 Centrality Analysis of the Five Major Ports in China

3.1 Overview of the Five Major Ports

China is located in the east of Asia, bordering the west coast of the Pacific Ocean, and its coastal ports are often hit by typhoons in summer and in this chapter, based on the Lloyd's List Top 100 Global Container Ports in 2023 and considering the location of China's five major port groups, five representative ports have been selected for the centrality analysis under the impact of typhoons, which include the world's No. 1 port of Shanghai, No. 4 port of Shenzhen, No. 5 port of Qingdao, No. 14 port of Xiamen and No. 35 port of Qinzhou.

The Port of Shanghai, located at the mouth of the Yangtze River, is an important maritime gateway connecting China to the world. In 2023, the port's container throughput exceeded 49 million TEUs. In terms of being affected by typhoons, the number of typhoons that made landfall in Shanghai between 1949 and 2023 was 10. Although Shanghai has suffered fewer frontal landfalls from typhoons, it is often affected by the outer ring circulation of typhoons that land in Fujian and Zhejiang, causing port disruptions.

Shenzhen Port is an important container hub in southern China, with a superior geographical location. In 2024, Shenzhen Port's container throughput once again exceeded 30 million TEUs. According to statistics from the China Weather Network, the number of typhoons that have made direct landfall in Shenzhen since 1949 is 13. However, Shenzhen Port is affected by typhoons more frequently than other ports.

The Port of Qingdao is a key hub for foreign trade and maritime transportation in northern China. In 2023, the Port of Qingdao's cargo throughput reached 664 million tons. Since 1949, only 23 typhoons have made landfall in Shandong Province, where Qingdao Port is located, with the maximum landfall intensity only reaching the Strong Tropical Storm (STS) level and the frequency and intensity of the impacts being the mildest among the five sample ports.

Xiamen Port, an important port on the southeast coast of China, is often affected by typhoons. Historically, typhoons such as Typhoon Dann in 1999 and Typhoon Meranti in 2016 have severely impacted Xiamen Port. Typhoons bring strong winds, heavy rains and storm surges, which pose a threat to port facilities and can lead to operational disruptions.

Qinzhou Port is an important seaport and maritime transportation hub in southwestern China, connecting over 200 ports in more than 100 countries and regions worldwide. In terms of typhoon impact, Qinzhou Port is located in the high typhoon area in the Northwest Pacific Ocean and the South China Sea region, and it often suffers from strong winds, heavy rains and other peripheral impacts.

3.2 *Port Centrality Weights Assessment Method in the Context of Typhoons*

The entropy weight method is based on the indicator's objective data to calculate the indicator's entropy value to judge its degree of discrete (He et al., 2016).

1. Construct the data matrix $(X_{ij})_{m\times n}$.

   The value of the jth indicator of the ith port is noted as X_{ij}, and the original data matrix is constructed with the following formula:

$$\left(X_{ij}\right)_{m\times n}=\begin{bmatrix} x_{11} & x_{12} & \cdots & x_{1n} \\ x_{21} & x_{22} & \cdots & x_{2n} \\ \cdots & \cdots & \cdots & \cdots \\ x_{m1} & x_{m2} & \cdots & x_{mn} \end{bmatrix} \quad (4.1)$$

$$\left(i=1,2,\ldots,\mathrm{m};j=1,2,\ldots,n\right)$$

2. Standardisation of data

   Standardise the raw data to eliminate the influence of different indicator units.
   Positive indicators:

$$y_{ij}=\frac{x_{ij}-\min\left(x_j\right)}{\max\left(x_j\right)-\min\left(x_j\right)} \quad (4.2)$$

Negative indicator:

$$y_{ij} = \frac{\max\left(x_j\right) - x_{ij}}{\max\left(x_j\right) - \min\left(x_j\right)} \tag{4.3}$$

where y_{ij} denotes the normalised value of the jth evaluation indicator of the ith port.

3. Leveling of Evaluation Indicators
   Normalise the original data and add 0.0001 to each value.
4. Calculation of the proportion of the ith port in the jth evaluation indicator

$$P_{ij} = \frac{x_{ij}}{\sum_{i=1}^{n} x_{ij}} \tag{4.4}$$

The normalisation matrix $(P_{ij})_{m \times n}$ is obtained by normalizing the indicators:

$$\left(P_{ij}\right)_{m\times n} = \begin{bmatrix} P_{11} & P_{12} & \cdots & P_{1n} \\ P_{21} & P_{22} & \cdots & P_{2n} \\ \cdots & \cdots & \cdots & \cdots \\ P_{m1} & P_{m2} & \cdots & P_{mn} \end{bmatrix} \tag{4.5}$$

5. Information entropy of the jth evaluation index e_j

$$e_j = -\frac{1}{\ln\left(n\right)} \sum_{j=1}^{n} P_{ij} \ln P_{ij} \tag{4.6}$$

where, $e_j \geq 0$.

6. Information utility value of evaluation index d_j

$$d_j = 1 - e_j \tag{4.7}$$

7. The value of weight coefficient of evaluation index w_j

$$w_j = \frac{d_j}{\sum_{j=1}^{m} d_j} \tag{4.8}$$

3.3 Centrality Analysis of Five Major Ports

The importance of ports includes their network centrality, economic impact, logistics role and connectivity. The port disruptions caused by typhoons are crucial for global trade and regional economies and can seriously affect shipping networks. For example, interruptions may slow down the transportation of goods, increase costs, affect the supply chain, reduce business efficiency, and thus have a wide-ranging impact on the shipping market.

In order to enhance the persuasiveness of assessing the importance of ports, this chapter refers to Alphaliner's Global Top 100 Shipping Companies Capacity Ranking (as of June 7, 2024). Data shows that the total shipping capacity of the top 12 liner companies reached 25.9188 million TEUs, accounting for 87.3% of the global market share. Based on this data, we collected route information from these 12 companies, totalling over 1200 routes covering 267 ports. We constructed a detailed maritime network using UCINET software to visualise and analyse the importance of sample ports. The five sample ports in the maritime network diagram show strong connectivity with many node links. In order to evaluate the importance of these ports more comprehensively, five different indicators, such as degree centrality, intermediary centrality and proximity centrality, are considered for analysis (Wan et al., 2021).

3.3.1 Degree Centrality

Degree centrality is an indicator for evaluating the importance of nodes in a network. If a port has more directly connected ports, its score will be higher, indicating that it is at the core of the maritime network.

$$DC = \sum_{j=1}^{n} a_{ij} \tag{4.9}$$

where a_{ij} is the number of connections between two port nodes i and j.

From the calculation results in Table 4.1, it can be seen that the five sample ports selected are all within the top twenty in the degree centrality ranking. Among them, the degree centrality value of Shanghai port is 78, the second highest among the five sample ports in the maritime network. This indicates that the port of Shanghai has a significant centrality in the maritime network and has extensive connections with other ports. In addition to ports in East Asia, ports in Southeast Asia occupy the majority of the top twenty positions. The reason for this phenomenon lies in the strategic location of the Southeast Asian region—it is not only the meeting point between Asia and Oceania but also connects the Indian and Pacific Oceans, thus constituting a key hub in the global shipping network. The Strait of Malacca, in particular, serves as an important passageway connecting East and West trade, further emphasizing the indispensable role of Southeast Asian ports in global shipping

Table 4.1 Results of port degree centrality

Port	Degree centrality	Rank	Port	Degree centrality	Rank
Singapore	124	1	Xiamen	41	11
Shanghai	78	2	Guangzhou	38	12
Shenzhen	73	3	Yokohama	36	13
Port Klang	70	4	Colombo	36	14
Busan	65	5	Piraeus	29	15
Tanjung Pelepas	57	6	Ho Chi Minh	29	16
HongKong	55	7	Jebel Ali	28	17
Ningbo	55	8	Tangier	27	18
Kaohsiung	54	9	Cai Mep	27	19
Qingdao	49	10	Qinzhou	26	20

routes. Therefore, compared to ports in other regions, Southeast Asian ports have more connecting routes and a higher centrality value. The high connectivity of these ports reflects their importance in the global shipping network and their key role in promoting international trade and regional economic development.

3.3.2 Betweenness Centrality

Betweenness Centrality measures the frequency of a node in the network as the other nodes to the shortest path to the intermediary point. Betweenness centrality reflects the ability of a node to act as a 'bridge' in the network, reflecting the node's control over the flow of information in the network. A node with high centrality is able to connect to other nodes more effectively, thus increasing its influence. For details of the calculation method, see Eq. 4.10.

$$\mathrm{BC} = \sum_{j=1}^{n} \frac{\sigma_{st}\left(i\right)}{\sigma_{st}} \tag{4.10}$$

where i is a node in the network; s and t are any two nodes in the network; $\sigma_{st}(i)$ denotes the number of shortest paths from node s to node t that pass through node i; and σ_{st} denotes the number of all shortest paths from node s to node t.

According to the results in Table 4.2, it can be seen that in terms of intermediary centrality of the maritime network, the ports in East and Southeast Asia exhibit relatively high values, which are more prominent compared to other regions. Among them, the port of Shanghai still ranks first among the sample ports, which indicates that the port of Shanghai acts as a key transhipment node in the maritime transportation network and has a greater influence in controlling the flow of goods and information, making it a central hub for a global exchange of goods and data.

Although the port of Qinzhou is slightly inferior to the port of Xiamen in terms of degree centrality value, its intermediary centrality value unexpectedly surpasses the latter. This phenomenon is likely to be closely related to its unique geographical

Table 4.2 Results of port intermediary centrality

Port	Betweenness centrality	Rank	Port	Betweenness centrality	Rank
Singapore	11541.612	1	Qinzhou	1027.371	11
Shanghai	4177.001	2	Piraeus	881.864	12
Port Klang	3163.103	3	Guangzhou	852.392	13
Shenzhen	2933.603	4	Yokohama	760.858	14
Busan	2066.244	5	Jebel Ali	711.963	15
Tanjung Pelepas	1867.104	6	Cai Mep	535.968	16
HongKong	1575.449	7	Rotterdam	524.574	17
Ningbo	1348.109	8	Colombo	507.860	18
Qingdao	1121.444	9	Xiamen	492.854	19
Kaohsiung	1082.723	10	Savannah	488.073	20

location. Qinzhou Port is located in the Beibu Gulf region of the South China Sea, and this strategic location makes it a material transit point for routes connecting Southeast and South Asia, greatly improving the efficiency of distributing goods through Qinzhou Port to ports in other regions and thus appearing to be more superior in terms of intermediary centrality.

3.3.3 Closeness Centrality

Closeness Centrality refers to the average distance of a node to all other nodes in the network. When a node's Closeness Centrality is higher, it means that it is more convenient in information transfer and resource acquisition.

$$CC = \frac{n-1}{\sum_{i \neq j} d_{ij}} \tag{4.11}$$

Where n is the number of all ports in the network; d_{ij} is the distance between node i to node j.

From the data in Table 4.3, it can be seen that the Shanghai and Shenzhen ports have the highest proximity centrality values, ranking second and third with 736 and 749. The high proximity centrality values of the ports of Shanghai and Shenzhen mean that they are able to quickly connect with other ports in the network and thus play a crucial role in the global shipping network.

The ports of Qingdao and Xiamen rank 10th and 11th, respectively, with proximity centrality values of 788 and 800. They are regional shipping centres and important ports for international trade. Qinzhou Port has an important strategic position on the southwest coast of China and is a key shipping node connecting China with ASEAN countries. Although slightly behind in the ranking, it is still located in the nineteenth position in the network,

The proximity centrality data of the five sample ports and their rankings in the network show that these five ports have strong connectivity and influence in the

Table 4.3 Results of port closeness centrality

Port	Closeness centrality	Rank	Port	Closeness centrality	Rank
Singapore	688	1	Xiamen	800	11
Shanghai	736	2	Guangzhou	802	12
Shenzhen	749	3	Cai Mep	813	13
Port Klang	749	4	Laem Chabang	821	14
Tanjung Pelepas	770	5	Colombo	821	15
Busan	773	6	Jebel Ali	825	16
Ningbo	776	7	Ho Chi Minh	828	17
HongKong	777	8	Fuzhou	833	18
Kaohsiung	779	9	Qinzhou	835	19
Qingdao	788	10	Taipei	836	20

global shipping network, which not only attracts more routes and shipping companies but also enhances the competitiveness of shipping in the whole region and provides a solid foundation for the smooth operation of global trade.

3.3.4 Eigenvector Centrality

Eigenvector centrality is a measure of node importance, taking into account the number and importance of connected nodes. In a maritime network, if a port is connected to a port with high eigenvector centrality, then the eigenvector centrality of that port will also increase. Thus it is more likely to become a key node for international trade and shipping, and its importance in the network will also increase. The eigenvector centrality formula is shown in Eq. 4.12.

$$\mathrm{EC}_i = \frac{1}{\lambda}\sum_{j=1}^{n} A_{j,i}\mathrm{EC}_j \tag{4.12}$$

Where λ is a constant; A is the adjacency matrix, $A_{j,i} = 1$ represents that node i and node j are connected, and $A_{j,i} = 0$ represents that node i and node j are not connected.

From the calculation results in Table 4.4, it can be seen that Singapore port eigenvector centrality tops the list in the network, which matches its international status as a global shipping hub. The ports of Shanghai and Shenzhen follow, and the difference in their eigenvector centrality values is not large, ranking second and third, respectively. This result shows that the other ports connected to these two ports are relatively balanced in terms of their degree of importance, reflecting the importance of the Shanghai and Shenzhen ports in the maritime transportation network and their close connection with other important ports. The Port of Shanghai, as one of the largest shipping centres in China, has been gaining influence and control in the global shipping network, while the Port of Shenzhen has taken its place in the global shipping network with its rapid growth and strong regional connectivity. The eigenvector values of Qingdao Port, Xiamen Port and Qinzhou Port are all in the top 20 of the network ranking, far exceeding the average value of 5.591, which proves that the five sample ports occupy an important position in the constructed maritime network.

This data and ranking strongly prove that these five ports occupy a significant position in the constructed maritime transportation network. Overall, these ports are strategically located and important in terms of their influence and connectivity in the maritime network.

3.3.5 Core-Edge Structure Analysis

Core-edge analysis is a method for identifying key nodes and secondary nodes in complex networks. Among them, the core nodes are generally located in the centre of the network with high centrality, while the edge nodes are relatively isolated, and the peripheral nodes are sparsely connected to each other, forming a scattering edge distribution, which has less influence on the whole network (Sciarra et al., 2018). The results of the port core degree and ranking of the maritime network are shown in Table 4.5.

Table 4.4 Results of port eigenvector centrality

Port	Eigenvector centrality	Rank	Port	Eigenvector centrality	Rank
Singapore	37.324	1	Xiamen	25.486	11
Shanghai	33.923	2	Guangzhou	22.186	12
Shenzhen	33.922	3	Laem Chabang	19.405	13
Busan	29.918	4	Ho Chi Minh	18.824	14
Ningbo	28.939	5	Cai Mep	18.583	15
Kaohsiung	28.884	6	Haiphong	18.383	16
Hong Kong	28.317	7	Manila	17.360	17
Port Klang	26.858	8	Yokohama	16.945	18
Qingdao	26.532	9	Taipei	16.919	19
Tanjung Pelepas	26.347	10	Qinzhou	14.812	20

Table 4.5 Results of port core degree

Port	Core degree	Rank	Port	Core degree	Rank
Singapore	0.275	1	Xiamen	0.181	11
Shanghai	0.246	2	Guangzhou	0.156	12
Shenzhen	0.246	3	Laem Chabang	0.136	13
Busan	0.214	4	Ho Chi Minh	0.132	14
Ningbo	0.207	5	Cai Mep	0.130	15
Kaohsiung	0.206	6	Haiphong	0.129	16
Hong Kong	0.202	7	Manila	0.122	17
Port Klang	0.191	8	Taipei	0.118	18
Qingdao	0.188	9	Yokohama	0.118	19
Tanjung Pelepas	0.187	10	Qinzhou	0.103	20

Through in-depth analysis of the comprehensive port core degree data, it can be concluded that the average core degree of 267 ports is 0.039. Using the average core degree as a criterion, a total of 86 ports have a core degree that exceeds this average, which belongs to the 'core ports' in the maritime network and plays a key role in the maritime network; the remaining 181 ports are regarded as 'core ports', and the remaining 181 ports are regarded as 'core ports', and the remaining 181 ports are regarded as core ports. The remaining 181 ports are regarded as 'marginal ports' and play a relatively small role in the network.

Further observation of the data in Table 4.5 also shows that the five sample ports selected in the article are all core ports, which play a key role in the constructed maritime transportation network. It is worth mentioning that Chinese ports occupy half of the top 20 ports in the core degree ranking, among which Shanghai Port and Shenzhen Port are among the top three, which fully demonstrates the important position of Chinese ports in this maritime network and highlights China's significant influence in the field of global trade and shipping.

In summary, this chapter firstly utilises UCINET software to draw a shipping network map including the five major ports in China, and then calculated five key indicators including degree centrality, intermediary centrality, etc. to measure the importance of the five major ports (Tables 4.6 and 4.7). Subsequently, the entropy weighting method and Python software were used to objectively weight the five measurement indicators in order to obtain a comprehensive assessment of the importance of the sample ports. Finally, the port comprehensive score formula was utilised to calculate the comprehensive score and ranking of the importance of each sample port. The port comprehensive score formula is shown in Eq. 4.13.

$$S_i = \sum_{j=1}^{5} w_j P_{ij} \tag{4.13}$$

Table 4.6 Objective weights for port centrality measurement indicators

	Degree centrality	Betweenness centrality	Closeness centrality	Eigenvector centrality	Core degree
Objective weight	20.15%	29.98%	18.43%	15.62%	15.83%

Table 4.7 Overall results of port centrality

Port	Centrality comprehensive score	Rank
Shanghai	1.0001	1
Shenzhen	0.8553	2
Qingdao	0.4177	3
Xiamen	0.2969	4
Qinzhou	0.0436	5

A careful analysis and comprehensive assessment of the centrality of the five major ports shows that the Port of Shanghai ranked first in the sample ports in terms of the composite score of importance, a result that matches the objective international status of the Port of Shanghai as an international shipping center as well as its degree of influence on the entire shipping network.

It is closely followed by Shenzhen Port, which ranks second with a composite score of 0.8553, just after Shanghai Port. Despite its second place, the gap between Shenzhen Port and Shanghai Port in terms of composite score is not large, showing that its importance is also quite significant and should not be underestimated.

In third place is the Port of Qingdao, with a composite score of 0.4177, which is only half the score of the first two ports, thus showing that the degree of importance of the ports of Shanghai and Shenzhen is much higher than that of the other sample ports. The fourth and fifth ports are Xiamen Port and Qinzhou Port, each showing a key role in regional shipping.

Overall, the ranking of ports in terms of the overall centrality score is broadly in line with the previously calculated ranking of the five key indicator scores, which not only further validates the reasonableness of the assessment results but also provides a clear perspective to understand the status and role of ports in the global shipping network, and helps to better grasp the dynamics and trends of global shipping.

4 Port Resilience Assessment

4.1 Indicator Framework for Evaluating Port Resilience in the Context of Typhoons

A generally accepted evaluation framework for port risk assessment has not yet been formed. Based on this, this chapter refers to relevant literature on port risk assessment and resilience research and establishes a port resilience assessment index system for the typhoon context. The system contains three primary indicators and 11 secondary indicators. The details of the framework are shown in Table 4.8.

All data comes from the official websites of five local ports and transportation bureaus, the Ministry of Transport of China, and the National Port Statistical Yearbook. All indicators belong to positive indicators. That is, the larger the indicator value, the stronger the corresponding capacity of the port. In the indicator, C31 is the equipment recovery capability, expressed as the annual GDP of the port city. C32 is the human recovery capability, expressed as a percentage of senior port workers.

Table 4.8 Indicator framework for evaluating port resilience under typhoon impacts[a]

Target layer	Primary indicator	Secondary indicators	Unit	Attribute	References
Assessment of Port Resilience under Typhoon Impact (A)	Resilience B1	Length of the breakwater in the work area C11	metre	+	Hossain et al. (2019); Jiang et al. (2021); Wang(2011)
		Number of loading and unloading equipment C12	per	+	
		Coverage rate of dock shore power facilities C13	%	+	
		Number of berths C14	per	+	
		Number of port network security assessments C15	per	+	
		Equipment maintenance and ship inspection frequency C16	per	+	
		Number of relevant training and emergency drills C17	per	+	
	Adaptive Capacity B2	Sea rail intermodal throughput C21	TEU	+	
		Number of anchorages C22	per	+	
	Restorative Capcity B3	Equipment recovery capability C31	hundred million yuan	+	
		Human recovery capability C32	%	+	

[a]Note: Indicator attributes reflect the correlation between the evaluation indicator and port resilience. When the indicator is positively related to resilience, i.e., the attribute is "+", it indicates that an increase in the value of the indicator will lead to an increase in port resilience; on the contrary, when the indicator is inversely related to resilience, i.e., the attribute is "-", it indicates that the larger the value of the indicator is, the less resilient the port will be

4.2 Port Resilience Assessment Model in the Context of Typhoons

4.2.1 Hierarchical Analysis

The Analytic Hierarchy Process (AHP) method was proposed in the early 1970s for complex situations such as multi-criteria and multi-attribute (Saaty, 2000). The main steps of the method include:

Before calculating the weights of indicators, it is necessary to establish a progressive hierarchical structure model and construct a judgement matrix. Divide complex decisions into multiple levels, forming a progressively layered structure. Then, based on the idea of the "consistent matrix method" proposed by Saaty (1980), the relative importance matrix A between factors is obtained by pairwise comparison.

The first step in calculating the weight of indicators is to normalise the judgement matrix A by column to obtain the standard judgement matrix B:

$$B_{ij} = \frac{A_{ij}}{\sum_{k=1}^{n} A_{kj}} \tag{4.14}$$

Then, sum the standard judgment matrix B_{ij} by rows to get the sum vector W_i:

$$W_i = \sum_{j=1}^{n} B_{ij} \tag{4.15}$$

Finally, the obtained sum vector W_i is normalised to obtain the weights $\overline{W}_i$:

$$\overline{W}_i = \frac{W_i}{\sum_{j=1}^{n} W_i} \tag{4.16}$$

According to the weights $\overline{W}_i$, the maximum characteristic roct $\lambda_{\max}$ of the judgment matrix A can be calculated and used for the consistency test in the next step:

$$\lambda_{\max} = \frac{1}{n}\sum_{i=1}^{n} \frac{A_i \overline{W_i}}{\overline{\overline{W_i}}} \tag{4.17}$$

After determining the weights of the indicators, the consistency test must be carried out to ensure the reasonableness of the matrix calculation results.

First, the consistency index C.I. (consistency index) is calculated:

$$\text{C.I.} = \frac{\lambda_{\max} - n}{n-1} \tag{4.18}$$

Then, based on the order of the judgment matrix, determine the corresponding average random consistency index R I. (random index).

Finally, the consistency indicator C.R. is calculated with the following formula:

$$\text{C.R.} = \frac{\text{C.I.}}{\text{R.I.}} \tag{4.19}$$

When C.R. < 0.1, it represents that the judgment matrix passes the consistency test; when C.R. > 0.1, it is necessary to adjust or re-construct the judgment matrix until the judgement matrix passes the consistency test.

4.2.2 Entropy Weight-AHP Combination Assignment

The entropy weight AHP method is used to obtain the comprehensive weight of evaluation indicators.

$$\theta_j = \frac{\bar{W}_i w_j}{\sum_{i,j=1}^{n} \bar{W}_i w_j} \tag{4.20}$$

4.2.3 TOPSIS Method

The TOPSIS method assesses evaluation objects' scores and rankings by calculating their distances to the best and worst index values [28].

Step 1: Calculate the weighted evaluation matrix $(Z_{ij})_{m \times n}$ of the indicators.

$$\left(Z_{ij}\right)_{m\times n} = \begin{bmatrix} \theta_1 P_{11} & \theta_1 P_{12} & \cdots & \theta_1 P_{1n} \\ \theta_2 P_{21} & \theta_2 P_{22} & \cdots & \theta_2 P_{2n} \\ \cdots & \cdots & \cdots & \cdots \\ \theta_n P_{m1} & \theta_n P_{m2} & \cdots & \theta_n P_{mn} \end{bmatrix} \tag{4.21}$$

Step 2: Calculate the positive ideal value Z+ and negative ideal value Z- in the weighted evaluation matrix.

$$Z^+ = \left(Z_1^+, Z_2^+, \cdots, Z_m^+\right) = \begin{pmatrix} \max\left\{\theta_1 P_{11}, \theta_2 P_{21}, \cdots, \theta_n P_{m1}\right\}, \\ \max\left\{\theta_1 P_{12}, \theta_2 P_{22}, \cdots, \theta_n P_{m2}\right\}, \\ \cdots, \\ \max\left\{\theta_1 P_{1n}, \theta_2 P_{2n}, \cdots, \theta_n P_{mn}\right\} \end{pmatrix} \tag{4.22}$$

$$Z^- = \left(Z_1^-, Z_2^-, \cdots, Z_m^-\right) = \begin{pmatrix} \min\left\{\theta_1 P_{11}, \theta_2 P_{21}, \cdots, \theta_n P_{m1}\right\}, \\ \min\left\{\theta_1 P_{12}, \theta_2 P_{22}, \cdots, \theta_n P_{m2}\right\}, \\ \cdots, \\ \min\left\{\theta_1 P_{1n}, \theta_2 P_{2n}, \cdots, \theta_n P_{mn}\right\} \end{pmatrix} \tag{4.23}$$

Step 3: Calculate the distance d_i of each evaluation object from the positive and negative ideal values.

$$d_i^+ = \sqrt{\left[\sum_{j=1}^{n}\left(Z_{ij} - Z_j^+\right)\right]^2} \quad (i = 1 2 \cdots \quad m) \tag{4.24}$$

$$d_i^- = \sqrt{\left[\sum_{j=1}^{n}\left(Z_{ij} - Z_j^-\right)\right]^2} \quad (i = 1 2 \cdots \quad m) \tag{4.25}$$

Step 4: Calculate the relative proximity of each evaluation object and obtain a score Ci. The higher the score, the higher the port resilience.

$$C_i = \frac{d_i^-}{d_i^+ + d_i^-} (i = 1 2 \cdots \quad m) \tag{4.26}$$

4.3 *Weights of Evaluation Indicators*

The subjective weight of resilience evaluation indicators was determined using a questionnaire survey method, and responses were collected from 30 experts in the shipping industry. The subjective weight was calculated using the AHP method. Objective weight is determined by collecting data from ports and using the entropy weight method. Finally, the two are combined to obtain comprehensive weights, and the TOPSIS method is used to obtain the relative closeness and ranking of Shanghai Port and Shenzhen Port with the ideal value of resilience (Tables 4.9 and 4.10).

From the calculation results, it can be seen that the relative proximity value of the Shanghai port is 0.6420 and the relative proximity value of the Shenzhen port is 0.3580, i.e., in the context of the typhoon, the resilience index of the Shanghai port is higher than that of the Shenzhen port, and the Shanghai port's overall ability to recover after being affected by the typhoon is slightly better than that of the Shenzhen port.

Table 4.9 Weights of evaluation indexes of the resilience of five major ports in the context of typhoon

	Subjective weight $^{-}W_i$	Objective weight W_j	Comprehensive weight θ_j
C11	0.0339	0.0909	0.0377
C12	0.0352	0.0909	0.0391
C13	0.0382	0.0909	0.0424
C14	0.0511	0.0909	0.0568
C15	0.0464	0.0909	0.0516
C16	0.0474	0.0909	0.0526
C17	0.0478	0.0909	0.0532
C21	0.1508	0.0909	0.1676
C22	0.1492	0.0909	0.1657
C31	0.1375	0.0909	0.1528
C32	0.1625	0.0909	0.1806

Table 4.10 Relative proximity and ranking of the sample ports to the ideal value

	Shanghai	Shenzhen
Relative approach degree C_i	0.6420	0.3580
Rank	1	2

5 Conclusion

The five major ports in China occupy a pivotal position in the global shipping network, among which the ports in Shanghai and Shenzhen are particularly important. Through the comprehensive evaluation of these ports' centrality and intermediary centrality, it can be found that Shanghai Port has led the global container ports for many years by virtue of its superior geographical location and efficient logistics system, and its comprehensive importance score ranks first among the sample ports. The Port of Shenzhen follows closely behind, demonstrating its important role in global trade and strong business growth momentum. The ports of Qingdao, Xiamen and Qinzhou, although slightly behind in the overall score, also demonstrate a key role in regional shipping.

In terms of port resilience assessment, the Shanghai port has a higher resilience index than the Shenzhen port in the context of typhoons, indicating that the Shanghai port has a stronger overall resilience after being affected by typhoons. This is due to Shanghai Port's excellent absorption and recovery capabilities. Although Shenzhen Port performs well in some indicators, it is slightly inferior to Shanghai Port in terms of overall resilience.

Overall, the top five ports are strategically important in terms of geographic location, influence and connectivity of the maritime transportation network. The efficient operation of these ports provides a solid foundation for the smooth operation of global trade and also plays a key role in promoting the development of a regional economy. In the future, as the global shipping network further develops and changes, the importance and resilience of these ports will continue to receive attention, and their position and role in global shipping will continue to be strengthened and enhanced.

References

Cho, H., & Park, H. (2017). Constructing resilience model of port infrastructure based on system dynamics. *Disaster Management, 7*(3), 245–254. https://doi.org/10.2495/safe-v7-n3-352-360

Comnews. (2024, September 2). *Shijieji haiyang gangkouqun zonghe zhishu baogao (2024) zhengshi fabu* [The Report on Comprehensive Index of World-class Marine Ports (2024) was officially released]. https://www.comnews.cn/content/2024-09/02/content_44096.html

Gao, Z., Wan, R., Ye, Q., et al. (2020). Typhoon disaster risk assessment based on emergy theory: A case study of Zhuhai City, Guangdong Province, China. *Sustainability, 12*(10), 4212. https://doi.org/10.3390/su12104212

He, Y., Guo, H., Jin, M., & Ren, P. (2016). A linguistic entropy weight method and its application in linguistic multi-attribute group decision making. *Nonlinear Dynamics, 84*, 399–404. https://doi.org/10.1007/s11071-015-2595-y

Hong, K., & Ji, M. Z. (2019). Risk assessment of Typhoon Disasters in Nansha Port area based on different probability. *Journal of Tropical Meteorology, 35*, 604–613.

Hossain, N. U. I., Nur, F., Hosseini, S., et al. (2019). A Bayesian network-based approach for modeling and assessing resilience: A case study of a full service deep water port. *Reliability Engineering & System Safety, 189*, 378–396. https://doi.org/10.1016/j.ress.2019.04.037

Hsieh, C.-H. (2014). Disaster risk assessment of ports based on the perspective of vulnerability. *Natural Hazards, 74*(2), 851–864. https://doi.org/10.1007/s11069-014-1214-4

Huang, Z., He, Z., Zhao, P., et al. (2024). The effects of tropical cyclone on the container shipping network: A case study of Typhoon Ma-on (2022). *Transportation Research Part D: Transport and Environment, 136*, 104449.

Jiang, M., Lu, J., Qu, Z., et al. (2021). Port vulnerability assessment from a supply chain perspective. *Ocean & Coastal Management, 213*, 105851. https://doi.org/10.1016/j.ocecoaman.2021.105851

Kim, S., Choi, S., & Kim, C. (2021). The framework for measuring port resilience in Korean port case. *Sustainability, 13*(21), 11883. https://doi.org/10.3390/su132111883

Lin, H., Zeng, W., Luo, J., et al. (2022). An analysis of port congestion alleviation strategy based on system dynamics. *Ocean & Coastal Management, 229*, 106336. https://doi.org/10.1016/j.ocecoaman.2022.106336

Liu, G., Yin, J., Song, S., et al. (2023). Risk estimation of typhoon disaster based on three-dimensional information diffusion method. *Journal of Marine Science and Engineering, 11*(5), 1080. https://doi.org/10.3390/jmse11051080

Ministry of Transport of the People's Republic of China. (2024). *2023nian jiaotong yunshu hangye fazhan tongji gongbao* [Statistical Bulletin on the Development of Transportation Industry in 2023]. Retrieved from https://xxgk.mot.gov.cn/2020/jigou/zhghs/202406/t20240614_4142419.html

Mou, N., Xu, H., Liu, Y., et al. (2024). Does the Chinese coastal ports disruption affect the reliability of the maritime network? Evidence from port importance and typhoon risk. *International Journal of Applied Earth Observation and Geoinformation, 129*, 103846. https://doi.org/10.1016/j.jag.2024.103846

Panahi, R., Gargari, N. S., Lau, Y., et al. (2022). Developing a resilience assessment model for critical infrastructures: The case of port in tackling the impacts posed by the Covid-19 pandemic. *Ocean & Coastal Management, 226*, 106240. https://doi.org/10.1016/j.ocecoaman.2022.106240

Saaty, T. L. (1980). *The analytic hierarchy process*. McGraw-Hill.

Saaty, T. L. (2000). A scaling method for priorities in hierarchical structures. *Journal of Mathematical Psychology, 15*(3), 234–281. https://doi.org/10.1016/0022-2496(77)90033-5

Sciarra, C., Chiarotti, G., Laio, F., et al. (2018). A change of perspective in network centrality. *Scientific Reports, 8*(1), 15269. https://doi.org/10.1038/s41598-018-33336-8

Shafieezadeh, A., & Burden, L. I. (2014). Scenario-based resilience assessment framework for critical infrastructure systems: Case study for seismic resilience of seaports. *Reliability Engineering & System Safety, 132*, 207–219. https://doi.org/10.1016/j.ress.2014.07.021

Tian, Z., Jiang, F. C., Zhang, Y., et al. (2024). Assessing seaport disruption under tropical cyclones using influence diagram and physics-based modeling. *Transportation Research Part D: Transport and Environment, 132*, 104237.

Wan, C., Zhao, Y., Zhang, D., et al. (2021). Identifying important ports in maritime container ship networks along the Maritime Silk Road. *Ocean & Coastal Management, 211*, 105738. https://doi.org/10.1016/j.ocecoaman.2021.105738

Wan, C., Yuan, J., Cao, D., et al. (2024). A fuzzy evidential reasoning-based model for evaluating resilience of ports to typhoons. *Transportation Research Part D: Transport and Environment, 104228*.

Wang, J. J. (2011). *Yanhai gangkou dianxing ziran zaihai fengxian fenxi yu pinggu [Analysis and assessment of typical natural disaster risks in coastal ports]*. East China Normal University.

Wang, Y., Yin, Y., & Song, L. (2022). Risk assessment of typhoon disaster chains in the Guangdong–Hong Kong–Macau Greater Bay Area, China. *Frontiers in Earth Science, 10*, 839733. https://doi.org/10.3389/feart.2022.839733

Wang, N., Wu, M., & Yuen, K. F. (2023). Assessment of port resilience using Bayesian network: A study of strategies to enhance readiness and response capacities. *Reliability Engineering & System Safety, 237*, 109394. https://doi.org/10.1016/j.ress.2023.109394

Wang, T., Ng, A. K. Y., Wang, J., et al. (2024). Adaptation planning of container ports in the context of typhoon risks: The case of Ningbo-Zhoushan port in China. *Ocean & Coastal Management, 257*, 107303. https://doi.org/10.1016/j.ocecoaman.2024.107303

Yang, Y. C., & Ge, Y. E. (2020). Adaptation strategies for port infrastructure and facilities under climate change at the Kaohsiung port. *Transport Policy, 97*. https://doi.org/10.1016/j.tranpol.2020.06.019

Yuan, J., Wan, C., Huang, J., & Wang, T. (2023, August). Developing risk reduction strategies of Typhoon disaster for ports from the perspective of resilience. In *2023 7th International Conference on Transportation Information and Safety (ICTIS)* (pp. 2143–2150). IEEE.

Rebuilding Resilient Maritime Transport and Supply Chain Networks: The Yangtze River Delta, China

1 Introduction

With the intensification of global climate change and the frequent occurrence of natural disasters, ports, as key nodes in the global supply chain, are facing increasingly severe challenges. Port resilience is the ability of ports to withstand, adapt to, and recover quickly from disruptive events in the face of extreme weather events and other sudden disasters (Shaw et al., 2017). Port resilience has become a focus of research and an urgent need in practice. The maritime supply chain serves as the lifeline of global trade, and ports undertake cargo consolidation, distribution, storage, and transhipment tasks as the supply chain's core link. Therefore, ports occupy an irreplaceable position in the global maritime supply chain and have become an important hub for importing and exporting goods and trade transactions. Especially in the context of globalization, maritime transport has become the main mode of transport for international logistics, with nearly 80% of the global trade volume of goods and nearly 70% of the value of the global supply chain being completed through maritime transport (Gülmez et al., 2023; UNCTAD, 2023).

With the growth of international trade and the complexity of maritime shipping networks, the efficiency and resilience of ports directly affect the fluidity of the global supply chain. If ports are shut down due to disasters, it may lead to global logistics stagnation, causing significant losses in transnational supply chains, affecting production and consumption, and even having a far-reaching impact on the global economy. Therefore, enhancing port resilience is not only related to the sustainable development of ports themselves but also a core element to ensure the stable operation of the global maritime supply chain.

The Yangtze River Delta (YRD) is one of China's most economically active regions, and its maritime transportation network and ports play a crucial role in supporting domestic and international trade. As the central hubs linking China to the global economy, ports such as Shanghai Port and Ningbo Zhoushan Port carry

Y.-y. Lau et al., *Maritime Transport and Supply Chain Resilience*,
https://doi.org/10.1007/978-3-032-07566-6_5

large-scale cargo throughput and play a key role in promoting the efficient operation of international supply chains. However, with the increase in extreme weather events (e.g., tropical cyclones, storm surges, floods, etc.), ports in the YRD region are facing greater risk challenges. Taking tropical cyclones as an example, they have direct and indirect negative impacts on the maritime transportation system. In terms of direct impacts, strong winds damage port infrastructure, and heavy rainfall soaks goods, causing economic losses. At the same time, the interruption of port operations causes ships to dock for shelter, leading to port congestion. In severe cases, ports may also face the risk of paralysis due to their own problems. In terms of indirect impacts, extreme meteorological disasters impact the port network, causing congestion and interruption and deteriorating the transportation environment. This prompts ships to choose safe ports to change transportation routes or enter windproof anchorages for shelter. However, either choice will prolong voyage time, increase the load on normal ports, and put them at risk of interruption or congestion.

Under the above background, this paper takes two typical extreme meteorological disasters of tropical cyclones and storm surges as examples, adopts the risk assessment method based on an index system, and mainly analyses the impact of extreme meteorological disasters on the infrastructure and operation of coastal ports in the Yangtze River Delta region, and quantify the risk level of extreme meteorological disasters in ports. This paper not only deeply analyses the resilience performance of ports in case of disasters and provides actionable insights for port operators but also proposes strategies to improve the resilience of coastal ports in the Yangtze River Delta region and provides an important guarantee for the safe operation of the whole supply chain.

2 Literature Review

By reviewing the relevant articles on the impact of extreme weather hazards on coastal ports, it is found that the existing research results mainly focus on two aspects: on the one hand, the assessment of extreme weather hazards in coastal ports, and on the other hand, the study of the resilience and climate adaptation of coastal ports and their transportation networks, and the provision of strategic advice for the enhancement of their resilience.

2.1 Assessment of Extreme Weather Hazards in Coastal Ports

In the risk assessment of extreme weather hazards in coastal ports, the assessment methods are diverse, including risk modeling based on GIS, indicator systems, scenario simulations, and probability. In many articles, GIS is generally utilized with other combined methods to map risk assessment. In the risk assessment based on the scenario simulation method, scholars simulate various disaster scenarios by collecting a large amount of meteorological disaster data or using existing databases and doing relevant research topics based on the meteorological data obtained in the simulation process.

Zhang et al. (2022) and Abdelgawad et al. (2021) took the wind, storm surge, wave, and so on during the typhoon as the research object, considered the coupling effects of various disasters, and applied MIKE21 software, hydrodynamic model and other methods to simulate the disaster situation; Tian et al. (2024) proposed a new framework based on influence diagrams and Bayesian networks to comprehensively analyze the potential interruption risks faced by seaports during tropical cyclones in order to study the impact of extreme meteorological disasters on the interruption time of seaports. Repetto et al. (2017) systematically introduced a GIS-based meteorological monitoring and forecasting system, which provides operational information for the remote control of ports and more detailed extreme meteorological information about the ports to provide more detailed warnings of extreme meteorological hazards.

In the risk assessment based on data and probabilistic methods, Chua et al. (2021) and Wei et al. (2019) start from the perspective of port vulnerability and further reflect the loss of ports by calculating the port risk probability. Scholars Camus et al. (2019) and Verschuur et al. (2020) even link the port‘s interruption operation time with the port risk probability, enriching the research in this field.

In risk assessment based on the index system method, scholars usually use different evaluation models to analyze the impact of disasters on ports. For example, Wan et al. (2024) use the ER model and Utility function to evaluate the resilience of ports under typhoon disasters; Wang et al. (2023) take the Bayesian network model as the main body to analyze port resilience.

2.2 Port Resilience and Climate Adaptation

In recent years, researchers have proposed diverse models and analytical frameworks to assess the resilience of ports. They mostly take the two natural disasters of extreme weather and epidemics as the background and propose corresponding resilience strategies based on the experimental results. In the context of extreme meteorological disasters, researchers mainly analyze the adaptive capacity of ports to climate. For example, Fernández-Pérez et al. (2024a, b) proposed a composite risk analysis method based on a high-resolution probabilistic framework, which applies multi-model prediction to port risk assessment under different emission scenarios. Yang et al. (2018) developed a fuzzy Bayesian analysis framework, which analyzed the case adaptation of 14 major ports in Greater China by constructing a multi-standard decision model and combining evidence inference method. In addition, Wang et al. (2024a, b) simulated the impact of extreme climate events (such as a tropical cyclone) on the port supply chain network through a complex network model. Gharehgozli et al. (2017) proposed a "tricky problem" framework, which starts from four phases—historical data, stakeholder management, dynamic strategy development, and flexible implementation—to construct a step-by-step resilience enhancement pathway for adapting to climate change. Puig et al. (2024) studied the adaptation strategies of large ports around the world and found that many ports have made significant progress in the areas of clean energy and shore power supply, but there are still deficiencies in the implementation of environmental protection policies and cross-regional cooperation. Jiang et al. (2021) proposed that, through fuzzy theory and evidence-based reasoning, the adaptation of ports to climate change can be improved through a combination of environmental protection policies and cross-regional cooperation. The vulnerability of ports to climate hazards can be quantified through a fuzzy theory and evidence-based reasoning framework. In addition to this, individual scholars have pinpointed port infrastructure as the main body to study climate adaptation in ports. For example, Fernández-Pérez et al. (2024a, b) proposed an adaptation assessment framework to analyze the adaptability and effectiveness of each adaptation measure under different scenarios and time frames, and Fang and Zio (2019) proposed a novel mathematical framework based on adaptive robust optimization (ARO) and applied it to the critical infrastructure of ports.

In the context of other natural disasters, Hossain et al. (2019) took all the disaster factors such as technical and economic factors into account and further analyzed the quantification of port resilience by applying different methods such as backward and forward propagation, sensitivity analysis, and information theory, and suggested that alternative routes and manpower restoration, etc., are the important factors to improve the resilience. Qin et al. (2023) constructed the elastic econometric model of port nodes, and analyzed the characteristics and spatial characteristics of the elasticity of port nodes. Gu et al. (2023) proposed an integrated method combining the best worst method and fuzzy TODIM method, and applied their method to three port congestion, labor shortage, and freight rate abnormal fluctuation scenarios. Li et al. (2024) took the port group as a research object, proposed a model combination of hypergraph-based modeling and agent-based simulation, and found the complex relationship between the port and the ship. Different from the research background of the above authors, Gu et al. (2024) and Liu et al. (2023) used the background of COVID-19 to analyze the elasticity changes of ports during the epidemic.

In conclusion, based on the existing articles on the impact of extreme meteorological disasters on ports, it can be clearly observed that few factors of both port facilities and operation conditions are considered in the model or methods adopted in the article. Secondly, most of the articles are limited to a single port when analyzing the risk of extreme meteorological disasters in coastal ports, so the model method or the selected indicators used have certain limitations.

Therefore, based on the shortcomings of existing research in this chapter, this chapter selects port-related disaster evaluation indicators, adopts the CRITIC objective weighting method, and applies the matter-element extension risk assessment model to conduct extreme meteorological disaster risk assessment on the major coastal ports in the Yangtze River Delta region, and provides an important theoretical framework and practical guidance for port managers and policymakers.

3 Risk Assessment Model for Coastal Ports in the Context of Extreme Weather

3.1 Construction of Indicator System

Disaster risk assessment quantitatively assesses the likelihood and consequences of disasters in the risk area. And natural disaster risk assessment refers to a series of processes of evaluating the vulnerability of disaster-causing factors and disaster-bearing bodies that may bring potential threats or harm to lives, properties, livelihoods, and human-dependent environments, etc., and then determining the nature of the risk, the scope and the loss (Miao & Wen, 2023).

In March 2014, the Intergovernmental Panel on Climate Change (IPCC) officially released the Fifth Assessment Working Group II report Climate Change 2014: Impacts, Adaptation, and Vulnerability (2014,WGII, Glossary), a report that also formally introduces exposure and vulnerability into the conceptual framework of risk assessment, arguing that disaster risk assessment is a comprehensive consideration of the three dimensions of hazard-causing factors, exposure, and vulnerability.

3.1.1 Hazard of Disaster-Causing Factors

The scope, frequency, and intensity of occurrence of disaster-causing factors such as typhoons, rainstorms, and earthquakes. The higher the frequency of occurrence of disaster-causing factors, the wider the range of disaster-causing factors, and the higher the intensity of disaster-causing factors, the higher the disaster-causing risk is.

This paper selects tropical cyclones and storm surges as indicators of disaster-causing factors. Reading the related literature, the maximum wind speed and the number of tropical cyclones are selected to quantify the hazard index of tropical cyclones, and the maximum storm surge and the ultra-warning tidal level are selected to measure the hazard index of storm surge. Their calculation formulas are as follows.

1. Tropical cyclones

$$T_{zc} = T_{yzc} + T_{qzc}$$

where T_{yzc} indicates the ratio of disaster-causing factors, the frequency of occurrence of different wind speed levels is selected as an indicator of the intensity and frequency of occurrence of disaster-causing factors. T_{qzc} indicates the number of times a disaster has been recorded in the statistics, i.e., the occurrence of a disaster in a certain area due to a tropical cyclone is recorded as 1 time, and the occurrence of no disaster is recorded as 0 times.

$$T_{yzc} = 0.01F_1 + 0.02F_2 + 0.03F_3 + 0.04F_4$$

2. Storm surge
   Calculation of storm surge disaster degree (Table 1)

$$H_d = I_s \times 0.4 + I_w \times 0.6$$

where, H_d is the storm surge disaster degree; I_s is the storm surge intensity level index; I_w is the high tide level exceeding the alert degree level index.

$$I_s = S_1 \times 20 + S_2 \times 16 + S_3 \times 12 + S_4 \times 8 + S_5 \times 4$$

where S_i denotes the number of stations with storm surge level i.

$$I_w = H_1 \times 20 + H_2 \times 15 + H_3 \times 10 + H_4 \times 5$$

where H_i denotes the number of stations with high tide level exceeding the alert level of class i.

3.1.2 Exposure of the Disaster Receptor

The number of disaster receptors such as people, buildings, infrastructures, economy, and resource environment within the possible influence range of disaster-causing factors. The higher the exposure degree of the disaster receptor, the more likely to cause a greater disaster loss. Here, the authors select the indicators that can reflect the infrastructure and operation scale of coastal ports, including the annual throughput of port cargo, container throughput, quay length, and the number of berths as indicators of disaster receptor exposure.

3.1.3 Vulnerability of Disaster Receptor

Expression of the state or performance of the receptor by the adverse effects of the disaster-causing factors tendency, sensitivity, and vulnerability, generally reflected in the size of the disaster-causing hazards and the degree of loss of the disaster receptor of the relationship between the degree of loss. The greater the vulnerability

Table 1 Tropical cyclone hazard evaluation index grades and classification standards

Coefficient	Wind speed scale
0.01	6 ~ 8(F_1)
0.02	8 ~ 10(F_2)
0.03	10 ~ 12(F_3)
0.04	12 ~ 14(F_4)

of the disaster receptor, the higher the degree of loss of it in the case of the same disaster-causing hazard.

Here, the proportion of foreign trade cargo, the number of anchorages, the competitiveness of international shipping centers, and the government's dominant capacity are selected as the indicators of vulnerability of disaster receptor through reading related literature. Among them, the government's dominant capacity is expressed by the GDP of the city where the port is located, and the competitiveness of the international shipping center reflects the port's importance, which is expressed by calculating the port's centrality, which is calculated in the following formula.

The quantification of this indicator is achieved by applying Ucinet software to establish a shipping network diagram of ports in the Yangtze River Delta. In this way, the degree centrality, closeness centrality, and betweenness centrality of ports in the Yangtze River Delta are calculated. The entropy weight method is applied to weight the three centralities. Finally, the comprehensive centrality of each port is calculated.

1. Degree centrality

$$DC_i = \frac{k_i}{N-1}$$

where, k_i denotes the number of existing nodes connected to the points of node i; N is the number of network nodes.

2. Betweenness centrality

$$BC_i = \frac{2}{(N-1)(N-2)} \sum_{s \neq i \neq t} \frac{g_{st}^i}{g_{st}}$$

where g_{st} denotes the number of shortest paths connecting s and t; g_{st}^i denotes the number of shortest paths connecting s and t passing through node i.

3. Closeness centrality

$$CC_i = \frac{N-1}{\sum_{j=1}^{N} d_{ij}}$$

where d_{ij} denotes the shortest distance from node i to node j.

3.1.4 Establishment of Indicator System

According to the three dimensions of the index system construction, strictly following seven principles such as the systematicness, typicality, scientificity, and dynamism of index construction, the index system for extreme meteorological disaster risk assessment of coastal container ports in China is constructed as shown in Fig. 1, and the data used for each index is explained and illustrated, as shown in Table 2.

3.2 Indicator Weighting Methods

CRITIC method is a better objective assignment method than entropy weight and standard deviation methods. It is based on the comparative strength of the evaluation indicators and the conflict between them to comprehensively measure the objective weight of the indicators. Considering the size of indicator variability while considering the correlation between the indicators, not the larger the number means the more important, completely using the data's objective attributes to carry out the scientific evaluation. The calculation steps of the CRITIC method are shown below.

1. Dimensionless processing

   Divide the above indicators into positive and negative terms, and normalize the acquired data as follows.

   Numerical processing of positive indicators

$$x_{positive} = \frac{x_j - x_{\min}}{x_{\max} - x_{\min}}$$

   Numerical processing of the negative indicator

$$x_{negative} = \frac{x_{\max} - x_j}{x_{\max} - x_{\min}}$$

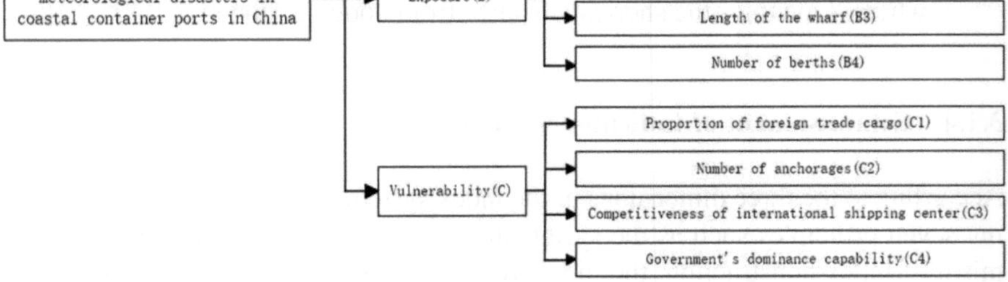

Fig. 1 Risk assessment index system of extreme meteorological disasters in coastal container ports

Table 2 Risk assessment indicators and data used

Index	Data	Reference
A1	Maximum wind speed (m/s) and tropical cyclone times are selected to quantify the risk index of tropical cyclones. The greater the value, the greater the risk.	Wang et al. (2022)
A2	Maximum storm surge (cm) and ultra-warning tidal level were selected to quantify the risk index of storm surge. The greater the value, the greater the risk.	Wang et al. (2022)
B1	The quantity of goods loaded and unloaded in the port within 1 year. The larger the value, the greater the exposure.	Wang et al. (2012)
B2	Number of containers handled and transported by the port within 1 year. The larger the value, the greater the exposure.	Wang et al. (2024a, b) and Wang et al. (2012)
B3	The actual length of ports used for berthing ships, loading and unloading cargo, and getting up and down passengers. The larger the value, the greater the exposure.	Chen (2022)
B4	The number of locations of ships that can be docked in the port area. The larger the value, the greater the exposure.	Chen (2022)
C1	The proportion of China's export goods in the total cargo throughput. The greater the value, the greater the vulnerability.	Wang et al. (2024a, b) and Wang et al. (2012)
C2	Waters with natural or artificial cover conditions against strong wind and waves. The greater the value, the smaller the vulnerability.	Wang et al. (2012)
C3	It reflects the shipping service of the port and represents the shipping service by calculating the importance of the port. The greater the value, the greater the vulnerability。	Wang et al. (2012)
C4	It reflects the disposable capacity of the city where the port is located, and mainly reflects the GDP of the port city. Greater magnitude of its value results in less vulnerability.	Wang et al. (2024a, b) and Wang et al. (2012)

2. Calculation of indicator variability

The standard deviation is used in the CRITIC method to indicate the fluctuation of the difference between the values taken within each indicator, i.e., the indicator variability. The larger the value of the standard deviation, the greater the difference in the value of the indicator, the more information can be reflected, the stronger the evaluation strength of the indicator itself, and more weight should be assigned to the indicator.

$$\overline{x}_j = \frac{1}{n}\sum\nolimits_{i=1}^{n} x_{ij}$$

$$S_j = \sqrt{\frac{\sum\nolimits_{i=1}^{n}\left(x_{ij} - \overline{x}_j\right)^2}{n-1}}$$

S_j denotes the standard deviation of the jth indicator

3. Calculation of indicator conflict

$$R_j = \sum_{i=1}^{p} \left(1 - r_{ij}\right)$$

   r_{ij} represents the correlation coefficient between i and j. In this paper, we use Pearson's correlation coefficient to calculate.
4. Calculation of the amount of information

$$C_j = S_j \sum_{i=1}^{p} \left(1 - r_{ij}\right) = S_j \times R_j$$

   The larger C_j is, the greater the role of the jth indicator in the whole evaluation index system, the more weight should be assigned to it.
5. Objective weight calculation

   Based on the above calculation, the objective weight of the j th indicator is

$$W_j = \frac{C_j}{\sum_{j=1}^{p} C_j}$$

3.3 *Extension Matter-Element Risk Assessment Model*

By determining the weights of the above indicators and collecting data, this subsection can assess the risk status of China's coastal ports, and this paper adopts the risk assessment model based on the theory of matter-element extension. Its calculation steps are as follows.

3.3.1 Determine the Object Element Matrix

The object element matrix is expressed as follows:

$$R = \begin{bmatrix} N_p & c_1 & v_1 \\ & c_2 & v_2 \\ & \vdots & \vdots \\ & c_n & v_n \end{bmatrix} = \begin{bmatrix} R_1 \\ R_2 \\ \vdots \\ R_n \end{bmatrix}$$

where R_i is then referred to as R for the partition element.

3.3.2 Determination of Classical and Nodal Domains

1. Determination of the classical domain.

   The classical domain material element R_j is denoted as:

$$R_j = \begin{bmatrix} N_j & c_1 & v_{j1} \\ & c_2 & v_{j2} \\ & \vdots & \vdots \\ & c_n & v_{jn} \end{bmatrix} = \begin{bmatrix} N_j & c_1 & (a_{j1}, b_{j1}) \\ & c_2 & (a_{j2}, b_{j2}) \\ & \vdots & \vdots \\ & c_n & (a_{jn}, b_{jn}) \end{bmatrix}$$

   where N_j denotes the jth grade of the object element, and v_{ji} denotes the range of values of the ith evaluation index of the object element under the feature grade.

2. Determination of the section domain.

   The section domain object element R_p is denoted as:

$$R_p = \begin{bmatrix} N_p & c_1 & v_{p1} \\ & c_2 & v_{p2} \\ & \vdots & \vdots \\ & c_n & v_{pn} \end{bmatrix} = \begin{bmatrix} N_p & c_1 & (a_{p1}, b_{p1}) \\ & c_2 & (a_{p2}, b_{p2}) \\ & \vdots & \vdots \\ & c_n & (a_{pn}, b_{pn}) \end{bmatrix}$$

   where v_{pi} denotes the range of values of the object element of the node domain with respect to the feature c_i, a_{pi}denotes the lower limit of the value interval of the ith metric, and b_{pi} denotes the upper limit of the value interval of the ith metric.

3.3.3 Determination of the Object Element to be Evaluated and the Correlation Function

1. Determination of object elements to be evaluated

$$R_x = \begin{bmatrix} N_x & c_1 & v_1 \\ & c_2 & v_2 \\ & \vdots & \vdots \\ & c_n & v_n \end{bmatrix}$$

2. Determination of the correlation function

$$k_{xj}\left(v_i\right)=\begin{cases}\dfrac{-\rho_{ij}\left(v_{xi},v_{ji}\right)}{\left|v_{ji}\right|} & \left(v_{xi}\in v_{ji}\right)\\[2ex] \dfrac{-\rho_{ij}\left(v_{xi},v_{ji}\right)}{\rho_{ij}\left(v_{xi},v_{pi}\right)-\rho_{ij}\left(v_{xi},v_{ji}\right)} & \left(v_{xi}\in v_{ji}\right)\end{cases}$$

where $k_{xj}(v_i)$ denotes the xth object element to be evaluated, the jth rank, and the correlation of the ith metric. $\rho_{ij}(v_{xi}, v_{ji})$ denotes the distance from a point x to the interval X, denoted by

$$\rho_{ij}\left(v_{xi},v_{ji}\right)=\rho_{ij}\left(x,X\right)=\left|x-\frac{1}{2}\left(a+b\right)\right|-\frac{1}{2}\left(b-a\right)$$

3.3.4 Determination of Integrated Relevance and Assessment Level

1. Determination of integrated relevance

$$K_j\left(N_x\right)=\sum\nolimits_{i=1}^{n}\omega_i k_i\left(v_i\right)$$

where, $\omega_i$indicates the comprehensive weight of the indicator obtained; $k_i(v_i)$ is the value of the correlation degree of the indicator obtained by the above calculation.

2. Determine the assessment level

$$K_j=\max\left(k_j\left(N_x\right)\right)$$

According to the principle of maximum affiliation, the risk level of extreme meteorological disasters in the assessed port can be determined by the above equation, and K_j represents the risk level of the study area.

4 Case Study

4.1 Introduction of the Ports in Yangtze River Delta

The Yangtze River Delta (YRD) is one of China's most important economic regions. It covers Shanghai, Jiangsu, Zhejiang, and part of Anhui, with a total area of 358,000 square kilometers. The region has rich port resources, including the Shanghai Port, Ningbo Zhoushan Port, and ports in Jiangsu. These ports play an indispensable role in domestic and international trade. Jiangsu has dense port resources, and its ports collaborate with the Shanghai Port. Zhejiang's Ningbo Zhoushan Port is an important connection to the global market. Although Anhui does not have a seaport, its inland ports provide support.

Although coastal container ports have more development opportunities than inland container ports, they also face greater risks of meteorological disasters. Therefore, this paper selects 15 coastal container ports in the Yangtze River Delta for extreme meteorological disaster risk assessment.

4.2 Data Sources

The data for the annual throughput of index cargo, annual container throughput, quay length, number of berths, and proportion of foreign trade cargo were sourced from the port yearbooks over the past 10 years. Information on anchorages is divided into two categories: data on anchorages at the mouth of the Yangtze River, obtained from the Ship E app; and data on anchorages at other ports, sourced from the website of China's Maritime Electronic Nautical Chart Online Service. Data on tropical cyclones were gathered from the Typhoon Network, while storm surge information was derived from China's Marine Disaster Bulletin and provincial marine disaster bulletins. Competitiveness data for international shipping centers were based on shipping routes operated by the top 12 global container liner companies, as reported by Alphaliner. Additional data were obtained from China's 2023 port yearbook, the official websites of each port, routes of shipping companies partnered with these ports, and local yearbooks of each port for information on governmental dominant capacity.

4.3 Calculation of Indicator Weights

According to the formulas in Sect. 3.2, the weights of each indicator layer can be obtained as shown in Table 3.

4.4 Risk Assessment of Major Ports in the Yangtze River Delta in the Context of Extreme Weather

According to the results of the weight assignment in the previous section, the major coastal ports in the Yangtze River Delta are assessed for the risk level of extreme meteorological disasters. The data after standardization of each indicator are arranged in order from the largest to the smallest, and the median method and the mean value method are used to classify the grades the risk level of extreme

Table 3 Determination of objective weights of indicators at each layer

First-level indicators	Objective weight	Second-level indicator	Objective weight
A	0.2783	A1	0.1551
		A2	0.1232
B	0.2478	B1	0.0512
		B2	0.0622
		B3	0.0597
		B4	0.0747
C	0.4739	C1	0.0840
		C2	0.1610
		C3	0.0706
		C4	0.1583

meteorological disasters is classified as low risk, general risk, large risk, and significant risk, which are expressed as Grade I, Grade II, Grade III, and Grade IV in that order. The criteria for classifying the grades of each indicator are shown in Table 4.

Based on Table 4, the classical domain $R_i$and the section domain R_p of the evaluation model are established as follows:

$$R_p = \begin{bmatrix} N_j & c_1 & (0\,1) \\ & c_2 & (0\,1) \\ & c_3 & (0\,1) \\ & c_4 & (0\,1) \\ & c_5 & (0\,1) \\ & c_6 & (0\,1) \\ & c_7 & (0\,1) \\ & c_8 & (0\,1) \\ & c_9 & (0\,1) \\ & c_{10} & (0\,1) \end{bmatrix}$$

Table 4 Grade classification of evaluation indicators

Index	Classical domain value interval division			
	I	II	III	IV
A1	[0,0.1088)	[0.1088,0.2300)	[0.2300,0.7449)	[0.7449,1]
A2	[0,0.1531)	[0.1531,0.7495)	[0.7495,0.8607)	[0.8607,1]
B1	[0,0.0551)	[0.0551,0.1855)	[0.1855,0.4092)	[0.4092,1]
B2	[0,0.0032)	[0.0032,0.0185)	[0.0185,0.3083)	[0.3083,1]
B3	[0,0.0518)	[0.0518,0.1369)	[0.1369,0.3912)	[0.3912,1]
B4	[0,0.0638)	[0.0638,0.1405)	[0.1405,0.4418)	[0.4418,1]
C1	[0,0.0798)	[0.0798,0.2041)	[0.2041,0.5217)	[0.5217,1]
C2	[0,0.6380)	[0.6380,0.8767)	[0.8767,0.9550)	[0.9550,1]
C3	[0,0.0011)	[0.0011,0.0028)	[0.0028,0.2179)	[0.2179,1]
C4	[0,0.6170)	[0. 6170,0.9010)	[0.9010,0.9496)	[0.9496,1]

$$R_1 = \begin{bmatrix} N_1 & c_1 & (0,0.1088) \\ & c_2 & (0,0.1531) \\ & c_3 & (0,0.0551) \\ & c_4 & (0,0.0032) \\ & c_5 & (0,0.0518) \\ & c_6 & (0,0.0638) \\ & c_7 & (0,0.0798) \\ & c_8 & (0,0.6380) \\ & c_9 & (0,0.0011) \\ & c_{10} & (0,0.6170) \end{bmatrix} \quad R_2 = \begin{bmatrix} N_2 & c_1 & (0.1088,0.2300) \\ & c_2 & (0.1531,0.7495) \\ & c_3 & (0.0551,0.1855) \\ & c_4 & (0.0032,0.0185) \\ & c_5 & (0.0518,0.1369) \\ & c_6 & (0.0638,0.1405) \\ & c_7 & (0.0798,0.2041) \\ & c_8 & (0.6380,0.8767) \\ & c_9 & (0.0011,0.0028) \\ & c_{10} & (0.6170,0.9010.) \end{bmatrix}$$

$$R_3 = \begin{bmatrix} N_3 & c_1 & (0.2300,0.7449) \\ & c_2 & (0.7495,0.8607) \\ & c_3 & (0.1855,0.4092) \\ & c_4 & (0.0185,0.3083) \\ & c_5 & (0.1369,0.3912) \\ & c_6 & (0.1405,0.4418) \\ & c_7 & (0.2041,0.5217) \\ & c_8 & (0.8767,0.9550) \\ & c_9 & (0.0028,0.2179) \\ & c_{10} & (0.9010,0.9496) \end{bmatrix} R_4 = \begin{bmatrix} N_4 & c_1 & (0.7449,1) \\ & c_2 & (0.8607,1) \\ & c_3 & (0.4092,1) \\ & c_4 & (0.3083,1) \\ & c_5 & (0.3912,1) \\ & c_6 & (0.4418,1) \\ & c_7 & (0.5217,1) \\ & c_8 & (0.9550,1) \\ & c_9 & (0.2179,1) \\ & c_{10} & (0.9496,1) \end{bmatrix}$$

Based on the above steps, the risk level of extreme meteorological hazards for major coastal ports in China‘s Yangtze River Delta region was calculated as shown in Table 5.

Table 5 Risk assessment results of extreme meteorological disasters in major coastal ports in the Yangtze River Delta

Comprehensive correlation degree	N1	N2	N3	N4	Risk grade
Lianyungang port	−0.4687	−0.4488	−0.3406	−0.5124	Greater risk
Nanjing port	−0.1919	−0.1725	−0.3379	−0.6931	Normal risk
Zhenjiang port	−0.2356	−0.4448	−0.3525	−0.5443	Low risk
Suzhou port	−0.2849	−0.5222	−0.1879	−0.5246	Greater risk
Yangzhou port	−0.2147	−0.4679	−0.5832	−0.6694	Low risk
Nantong port	−0.3741	−0.1458	−0.1462	−0.6102	Normal risk
Wuxi port	−0.3820	−0.1938	−0.3907	−0.6286	Normal risk
Taizhou Port (JS)	−0.3780	−0.2333	−0.4678	−0.5871	Normal risk
Changzhou port	−0.2830	−0.5682	−0.5707	−0.6447	Low risk
Yancheng port	−0.4091	−0.0425	−0.1899	−0.6573	Normal risk
Shanghai port	−0.5548	−0.6077	−0.4055	−0.2087	Significant risk
Jiaxing port	−0.4531	−0.2864	−0.0986	−0.5248	Greater risk
Ningbo-Zhoushan port	−0.6910	−0.7543	−0.8061	−0.1170	Significant risk
Taizhou port(ZJ)	−0.4743	−0.2303	−0.3690	−0.4161	Normal risk
Wenzhou port	−0.4021	−0.2098	−0.2831	−0.4502	Normal risk

(Note: JS for Jiangsu Province and ZJ for Zhejiang Province)

4.5 Risk Analysis of Extreme Meteorological Hazards in Major Coastal Ports in Yangtze River Delta

According to the risk assessment results in the previous section, it can be seen that there are three major coastal ports in China's Yangtze River Delta at low risk, seven general risk ports, three larger risk ports and two major risk ports. Among them, the major risk ports are Shanghai Port and Ningbo-Zhoushan Port, and the roles played by these two ports are crucial in China and even globally.

4.5.1 Hazard Analysis

A radar chart is drawn based on the correlation calculation results of each port hazard index, as shown in Fig. 2. Due to its geographical location and climatic conditions, the Yangtze River Delta region has a fairly high frequency and number of tropical cyclones and storm surges, and the degree of disaster impact is extremely serious. Major-risk ports include Shanghai Port, Ningbo-Zhoushan Port, Taizhou Port, and Wenzhou Port. Relatively high-risk ports include Suzhou Port, Nantong Port, and Jiaxing Port. General-risk ports include Lianyungang Port, Nanjing Port, Zhenjiang Port, Yangzhou Port, Wuxi Port, Taizhou Port, Changzhou Port, and Yancheng Port. From the perspective of the cities where they are located, Zhejiang

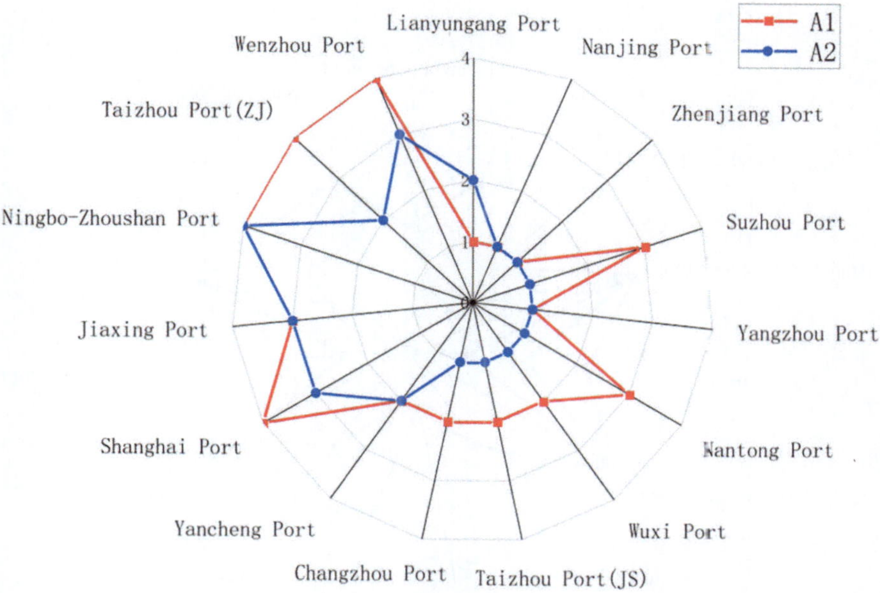

Fig. 2 Risk analysis of extreme meteorological disasters in major coastal ports of the Yangtze River Delta

Province is the city most severely affected by extreme meteorological disasters in the entire Yangtze River Delta region. Shanghai is second. Although Jiangsu has a high density of ports, most of them are far from the landing site of tropical cyclones, so the disaster level and degree are smaller than the first two.

4.5.2 Exposure Analysis

Based on the correlation calculation results of the various indicators of port exposure, a radar diagram was plotted as shown in Fig. 3. Among the major coastal ports in the Yangtze River Delta, Suzhou Port, Shanghai Port, and Ningbo-Zhoushan Port are major-risk ports. Changzhou Port is a low-risk port. Taizhou Port, Yangzhou Port, and Yancheng Port are general-risk ports. The rest are relatively high-risk ports. In general, the average exposure level of the port group in the Yangtze River Delta is at a relatively high risk. This is mainly because the port group in the Yangtze River Delta is the intersection of the north-south transportation corridors and is densely distributed. Moreover, river-sea combined transportation drives the connections between ports, promotes collaborative development, and the throughput reaches new highs year after year.

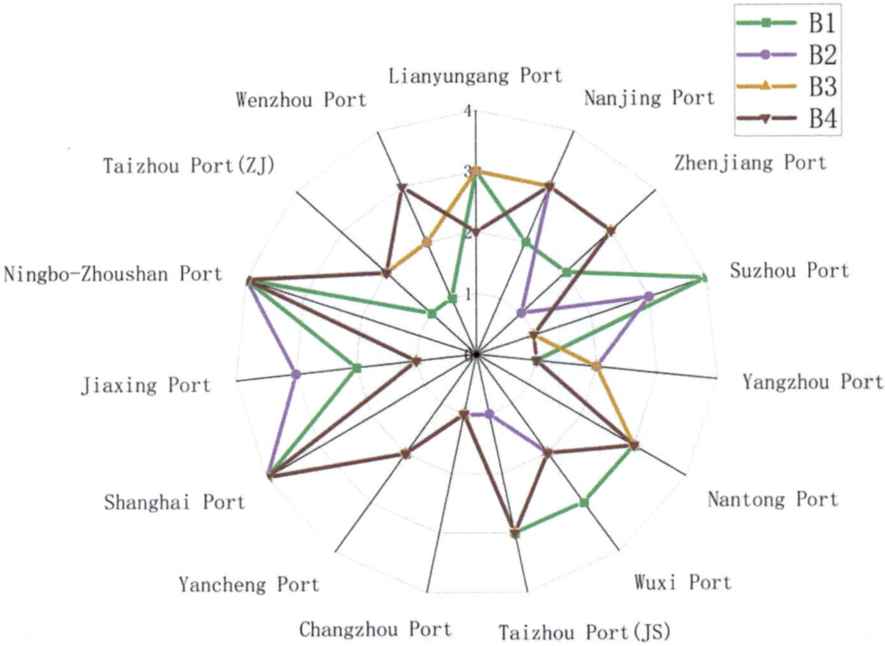

Fig. 3 Exposure analysis of extreme meteorological disasters in major coastal ports in the Yangtze River Delta

4.5.3 Vulnerability Analysis

A radar chart is plotted based on the correlation calculation results of each indicator of port vulnerability, as shown in Fig. 4. In the major coastal ports of the Yangtze River Delta, major-risk ports include Lianyungang Port, Zhenjiang Port, Yangzhou Port, Wuxi Port, Taizhou Port, Changzhou Port, Shanghai Port, Ningbo-Zhoushan Port, and Taizhou Port. Relatively high-risk ports are Suzhou Port, Nantong Port, Yancheng Port, Jiaxing Port, and Wenzhou Port. General-risk port is Nanjing Port. The ports downstream of the Yangtze River are at major risk because there are many ports, but the number of anchorages is limited, weakening their disaster prevention and reduction capabilities and making them more vulnerable to disasters. Secondly, Shanghai Port and Ningbo-Zhoushan Port are very important, so when affected, they endanger closely related ports, with a wide disaster impact range and greater losses and severer impacts.

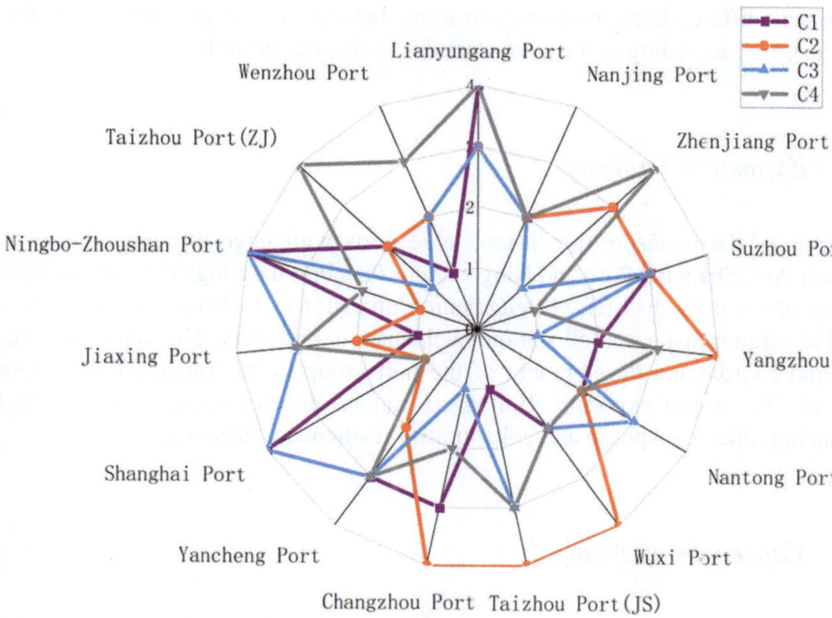

Fig. 4 Vulnerability analysis of extreme meteorological disasters in major coastal ports of the Yangtze River Delta

5 Conclusion and Future Implications

The assessment shows that the risk levels of major coastal ports in the Yangtze River Delta are diverse. There are three low-risk ports, seven average-risk ports, three high-risk ports, and two major-risk ports (Shanghai Port and Ningbo-Zhoushan Port). This distribution pattern shows the complexity and unevenness of the regional port system in the face of extreme weather disasters, and the resilience strategies required by ports of different risk levels to cope with disasters are different, and the major risk ports, because of their key position and high vulnerability, have become the key breakthrough point and focus of attention to improve the resilience of ports.

5.1 *Hazard Analysis*

From the hazard dimension, due to its geographical location and climatic conditions, the Yangtze River Delta experiences frequent tropical cyclones and storm surges. Ports in Zhejiang are the most severely affected, followed by Shanghai, and those in Jiangsu are relatively less affected. Ports in different regions should implement differentiated infrastructure strengthening and risk prevention and control measures according to the characteristics of disaster hazards.

5.2 Exposure Analysis

In terms of the exposure dimension, large ports with large throughput such as Suzhou Port, Shanghai Port, and Ningbo-Zhoushan Port have high exposure and are widely affected by disasters. Strengthening operational resilience and building a flexible cargo transfer and allocation mechanism is necessary. The dense distribution and coordinated development of the port group in the Yangtze River Delta increase the overall exposure. Building a unified disaster response system and enhancing linkage response and collaborative resilience is necessary.

5.3 Vulnerability Analysis

From the vulnerability dimension, the anchorage areas of ports downstream of the Yangtze River are in short supply, which restricts the ability to prevent and reduce disasters. Anchorage resources should be scientifically planned to improve the port‘s capacity to accommodate ships and conduct emergency dispatch. Core ports such as Shanghai Port and Ningbo-Zhoushan Port are widely affected by disasters and need to improve resilience in all aspects, build an emergency support system, play a leading role in core radiation, and improve the overall resilience and stability of the regional port network. In short, ports at different risk levels should take corresponding measures according to their own characteristics to improve port resilience and ensure the stable operation of the regional port group during disasters.

5.4 Implications for Enhancing Port Resilience

5.4.1 Strengthen Infrastructure Through Multiple Measures

Major-risk ports should upgrade disaster prevention facilities, such as raising embankments, improving drainage systems, and expanding anchorage areas. High-exposure ports need to add cargo protection facilities and efficient material transfer equipment. For ports with tight anchorage resources, optimizing berth resource allocation, strengthening the sharing and cooperation of anchorage resources among ports, and establishing a regional coordination management mechanism are necessary. At the same time, important ports should focus on improving shipping competitiveness and post-disaster recovery capabilities.

5.4.2 Build a Regional Collaborative Emergency Response System

The Yangtze River Delta should establish a unified disaster early warning system, integrate multiple types of monitoring data, accurately predict disasters, and issue early warnings through multiple channels. Formulate a regional linkage emergency response plan, optimize the allocation of emergency resources, clarify port responsibilities, establish a resource sharing and mutual assistance mechanism, achieve efficient linkage, and improve the overall emergency response speed and degree of coordination in the region are some examples on how to build a regional collaborative emergency response system.

5.4.3 Promote Green Intelligent Innovation and Strengthen Policy and Financial Support

There is a need to actively promote clean energy, build a port-distributed energy system, introduce intelligent drainage systems and real-time monitoring equipment, use big data and artificial intelligence technology to drive decision-making, and improve port operational efficiency and resilience. Establish a disaster risk fund to raise funds through multiple channels. The government should improve the policy incentive and restraint mechanism, support port enterprises adopting green and intelligent technologies, and strengthen the policy guarantee system to improve port resilience. These measures comprehensively improve the resilience of ports in the Yangtze River Delta and ensure stable operation during disasters.

References

Abdelgawad, M. A., Ellingwood, B., & Mahmoud, H. (2021). Vulnerability of seaports to hurricanes and sea—Level rise in a changing climate: A case study for mobile, AL. *Coastal Engineering, 167*, 103884. https://doi.org/10.1016/j.coastaleng.2021.103884

Camus, P., Tomás, A., Díaz-Hernández, G., Rodríguez, B., Izaguirre, C., & Losada, I. J. (2019). Probabilistic assessment of port operation downtimes under climate change. *Coastal Engineering, 147*, 12–24. https://doi.org/10.1016/j.coastaleng.2019.01.007

Chen, Z. Y. (2022). Port logistics function evaluation model based on entropy weight TOPSIS method. *Discrete Dynamics in Nature and Society, 2022*, 5006900. https://doi.org/10.1155/2022/5006900

Chua, C. T., Switzer, A. D., Suppasri, A., Li, L. L., Pakoksung, K., Lallemant, D., Jenkins, S. F., Charvet, I., Chua, T., & Cheong, A. (2021). Tsunami damage to ports: Cataloguing damage to create fragility functions from the 2011 Tohoku event. *Natural Hazards and Earth System Sciences, 21*(6), 1887–1908. https://doi.org/10.5194/nhess-21-1887-2021

Fang, Y. P., & Zio, E. (2019). An adaptive robust framework for the optimization of the resilience of interdependent infrastructures under natural hazards. *European Journal of Operational Research, 276*(3), 1119–1136. https://doi.org/10.1016/j.ejor.2019.01.052

Fernández-Pérez, A., Lara, J. L., Lucio, D., & Losada, I. J. (2024a). Compound climate change risk analysis for port infrastructures. *Coastal Engineering, 193*, 104560. https://doi.org/10.1016/j.coastaleng.2024.104560

Fernández-Pérez, A., Losada, I. J., & Lara, J. L. (2024b). A framework for climate change adaptation of port infrastructures. *Coastal Engineering, 191*, 104538. https://doi.org/10.1016/j.coastaleng.2024.104538

Gharehgozli, A. H., Mileski, J., Adams, A., & von Zharen, W. (2017). Evaluating a "wicked problem": A conceptual framework on seaport resilience in the event of weather disruptions. *Technological Forecasting and Social Change, 121*, 65–75. https://doi.org/10.1016/j.techfore.2016.11.006

Gu, B. M., Liu, J. G., & Chen, J. H. (2023). Scenario - based strategies evaluation for the maritime supply chain resilience. *Transportation Research Part D: Transport and Environment, 124*, 103948. https://doi.org/10.1016/j.trd.2023.103948

Gu, B. M., Liu, J. G., Ye, X. H., Gong, Y., & Chen, J. H. (2024). Data—Driven approach for port resilience evaluation. *Transportation Research Part E: Logistics and Transportation Review, 186*, 103570. https://doi.org/10.1016/j.tre.2024.103570

Gülmez, S., Sakar, G. D., & Bastug, S. (2023). An overview of maritime logistics: Trends and research agenda. *Maritime Policy & Management, 50*(1), 97–116. https://doi.org/10.1080/03088839.2021.1962557

Hossain, N. U. I., Nur, F., Hosseini, S., Jaradat, R., Marufuzzaman, M., & Puryear, S. M. (2019). A Bayesian network—Based approach for modeling and assessing resilience: A case study of a full—Service deep—Water port. *Reliability Engineering & System Safety, 189*, 378–396. https://doi.org/10.1016/j.ress.2019.04.037

IPCC. (2014). Summary for policymakers In: Climate change 2014: Impacts, adaptation, and vulnerability. part A: Global and sectoral aspects. In C. B. Field et al. (Eds.), *Contribution of Working Group II to the Fifth Assessment Report of the Intergovernmental Panel on Climate Change*(Cambridge Cambridge, United Kingdom and New York, NY, USA) (pp. 1–32). Cambridge University Press.

Jiang, M. Z., Lu, J., Qu, Z. H., & Yang, Z. L. (2021). Port vulnerability assessment from a supply chain perspective. *Ocean & Coastal Management, 213*, 105851. https://doi.org/10.1016/j.ocecoaman.2021.105851

Li, L. Y., Wei, C. Z., Liu, J., Chen, J. D., & Yuan, H. P. (2024). Assessing port cluster resilience: Integrating hypergraph—Based modeling and agent—Based simulation. *Transportation Research Part D: Transport and Environment, 136*, 104459. https://doi.org/10.1016/j.trd.2024.104459

Liu, J., Qi, Y., & Lyu, W. (2023). Port resilience in the post—COVID - 19 era. *Ocean & Coastal Management, 238*, 106565. https://doi.org/10.1016/j.ocecoaman.2023.106565

Miao, B., & Wen, J. (2023). A review of integrated risk assessment for natural disasters. *Earth Science Frontiers, 13*(8), 916–924. https://doi.org/10.12677/ag.2023.138087

Puig, M., Cirera, A., Wooldridge, C., Sakellariadou, F., & Darbra, R. M. (2024). Mega ports' mitigation, response and adaptation to climate change. *Journal of Marine Science and Engineering, 12*(7), 1112. https://doi.org/10.3390/jmse12071112

Qin, Y. F., Guo, J. K., Liang, M. X., & Feng, T. Q. (2023). Resilience characteristics of port nodes from the perspective of shipping network: Empirical evidence from China. *Ocean & Coastal Management, 237*, 106531. https://doi.org/10.1016/j.ocecoaman.2023.106531

Repetto, M. P., Burlando, M., Solari, G., De Gaetano, P., & Pizzo, M. (2017). Integrated tools for improving the resilience of seaports under extreme wind events. *Sustainable Cities and Society, 32*, 277–294. https://doi.org/10.1016/j.scs.2017.03.022

Shaw, D. R., Grainger, A., & Achuthan, K. (2017). Multi—level port resilience planning in the UK: how can information sharing be made easier? *Technological Forecasting and Social Change, 121*, 126–138. https://doi.org/10.1016/j.techfore.2016.10.065

Tian, Z., Jiang, F., Zhang, Y., Mahmoud, H., Lu, X. Z., Luo, M., Guo, J., & Guo, W. (2024). Assessing seaport disruption under tropical cyclones using influence diagram and physics—Based modeling. *Transportation Research Part D: Transport and Environment, 132*, 104237. https://doi.org/10.1016/j.trd.2024.104237

UNCTAD. (2023). Review of Maritime Transport (2023). https://unctad.org/system/files/official-document/rmt2023_en.pdf

Verschuur, J., Koks, E. E., & Hall, J. W. (2020). Port disruptions due to natural disasters: Insights into port and logistics resilience. *Transportation Research Part D: Transport and Environment, 85*, 102393. https://doi.org/10.1016/j.trd.2020.102393

Wan, C. P., Yuan, J. P., Cao, D. S., & Ng, A. K. Y. (2024). A fuzzy evidential reasoning—based model for evaluating resilience of ports to typhoons. *Transportation Research Part D: Transport and Environment, 133*, 104228. https://doi.org/10.1016/j.trd.2024.104228

Wang, J., Liu, M., Quan, R., Lu, M., Niu, H., & Xu, S. (2012). Natural disaster risk assessment of coastal ports in China. *Scientia Geographica Sinica, 32*(4), 516–520.

Wang, Y. J., Yin, Y. Z., & Song, L. C. (2022). Risk assessment of typhoon disaster chains in the Guangdong–Hong Kong–Macau Greater Bay Area, China. Frontiers. *Earth Science, 10*, 839733. https://doi.org/10.3389/feart.2022.839733

Wang, N. X., Wu, M., & Yuen, K. F. (2023). Assessment of port resilience using Bayesian network: A study of strategies to enhance readiness and response capacities. *Reliability Engineering & System Safety, 237*, 109394. https://doi.org/10.1016/j.ress.2023.109394

Wang, T. N., Feng, H. C., & Lau, Y. Y. (2024a). Analysis of the network efficiency of Chinese ports in global shipping under the impacts of typhoons. *Sustainability, 16*(8), 3190. https://doi.org/10.3390/su16083190

Wang, T. N., Ng, A. K. Y., Wang, J., Chen, Q., Pang, J. Y., & Tang, J. Q. (2024b). Adaptation planning of container ports in the context of typhoon risks: The case of Ningbo—Zhoushan port in China. *Ocean & Coastal Management, 257*, 07303. https://doi.org/10.1016/j.ocecoaman.2024.107303

Yang, Z. L., Ng, A. K. Y., Lee, P. T. W., Wang, T. N., Qu, Z. H., Rodrigues, V. S., Pettit, S., Harris, I., Zhang, D., & Lau, Y. Y. (2018). Risk and cost evaluation of port adaptation measures to climate change impacts. *Transportation Research Part D: Transport and Environment, 61*(B), 444–458. https://doi.org/10.1016/j.trd.2017.03.004

Zhang, Y., Zhang, X., Qiu, J., et al. (2022). A method for estimating the risk of fishery ports against typhoon: A case study on Dongsha fishery port. *Stochastic Environmental Research and Risk Assessment, 36*, 1993–2013. https://doi.org/10.1007/s00477-022-02243-w

Rebuilding Resilient Maritime Transport and Supply Chain Networks: ASEAN Countries

1 Introduction

The maritime industry is central to the economic vitality of the ASEAN region, a grouping of ten Southeast Asian nations strategically located at the crossroads of global trade routes (Lee, 2018). With an extensive network of ports and shipping lanes, ASEAN countries are integral to the flow of goods, passengers and resources worldwide (Carruthers et al., 2003). The region's ports, ranging from container hubs to cruise terminals, serve as critical nodes in the global supply chain, facilitating trade and tourism on an unprecedented scale.

However, this region is not immune to disruptions. The COVID-19 pandemic exposed vulnerabilities in maritime operations, causing delays and highlighting gaps in health protocols and logistics management (Panahi et al., 2022). Geopolitical shifts, such as the decoupling of China and the USA, have redefined trade dynamics, creating opportunities and challenges for ASEAN ports to reposition themselves (Rozman, 2022). Moreover, climate change—manifesting as rising sea levels, extreme weather events, and coastal erosion—threatens port infrastructure and the reliability of shipping lanes (Poo et al., 2021; Yang et al., 2018). These factors underscore the urgent need for resilience in maritime transport and supply chain networks.

This chapter explores the importance of maritime resilience in ASEAN countries through several lenses. It begins by introducing critical ports in the region, such as Singapore's globally renowned transhipment hub and emerging ports like Sihanoukville in Cambodia. Vulnerability assessments are highlighted, focusing on cruise port networks as a case study to demonstrate transferable methodologies for other systems. These assessments leverage tools like social network analysis (SNA) to identify network-level and port-level vulnerabilities, emphasising the interconnected nature of maritime operations. This chapter provides a comprehensive overview and actionable insights, aiming to contribute to the ongoing discourse on

Y.-y. Lau et al., *Maritime Transport and Supply Chain Resilience*,
https://doi.org/10.1007/978-3-032-07566-6_6

sustainable and resilient maritime systems in the ASEAN region and ensure that the industry remains robust in the face of global uncertainties.

2 Ports in ASEAN Countries

ASEAN countries are home to a diverse range of ports that cater to containerised cargo, bulk goods, and passenger movements (Yeo, 2024). These ports facilitate regional and international trade and serve as vital economic hubs. Below is an introduction to some of the key ports in the region.

2.1 Singapore

Singapore's port is one of the world's busiest and most advanced transhipment hubs, handling over 36 million twenty-foot equivalent units (TEUs) annually as of 2022 (Wan et al., 2021). The port's state-of-the-art facilities include automated cranes, digitised customs processing, and AI-powered logistics solutions. The Tuas Port project, set for completion by 2040, aims to consolidate Singapore's port operations into a single mega-hub capable of handling 65 million TEUs annually (Idris, 2022). The Marina Bay Cruise Centre has also grown significantly, welcoming over 1.8 million cruise passengers in 2019, compared to just one million in 2010 (Handani et al., 2022). This growth underscores Singapore's position as a cargo and passenger logistics leader.

2.2 Port Klang, Malaysia

Located near Kuala Lumpur, Port Klang is Malaysia's premier maritime gateway. Its three terminals—Northport, Southport, and Westport—collectively handled 13.2 million TEUs in 2022, up from seven million TEUs a decade earlier. This growth reflects increased trade activity driven by Malaysia's export-oriented industries (Mohd Rozar et al., 2023). Cruise operations at Port Klang have also expanded, with the terminal accommodating over 150,000 passengers annually before the pandemic. Significant investments in digitisation and hinterland connectivity have strengthened its regional competitiveness.

2.3 Tanjung Priok, Indonesia

As Indonesia's largest port, Tanjung Priok handled eight million TEUs in 2022, nearly doubling its throughput from 4.5 million TEUs in 2012 (Rachman & Wahid, 2023). The port has benefited from initiatives like the Jakarta Port Optimization Project, which introduced automated systems and expanded its container yard (Verhaeghe et al., 2021). Tanjung Priok also plays a role in passenger transport, hosting thousands of travellers annually on regional ferry networks. Its strategic location near Jakarta's industrial zones enhances its role as a logistics hub.

2.4 Laem Chabang, Thailand

Laem Chabang, located 130 kilometres from Bangkok, is Thailand's largest deep-sea port and a cornerstone of its export economy (Sukhotu, 2021). It handles 8.6 million TEUs in 2022, up from 5.3 million TEUs in 2012, and the port supports the automotive and electronics industries. Its Phase III expansion project includes new terminals and rail connections, aiming to boost capacity to 18 million TEUs annually by 2030. Cruise passenger arrivals peaked at 200,000 annually pre-pandemic, highlighting its growing appeal as a tourism hub (Ayuttacorn, 2023).

2.5 Ho Chi Minh City, Vietnam

The Cat Lai Terminal in Ho Chi Minh City is Vietnam's busiest container port, with a throughput of 6.3 million TEUs in 2022, doubling its volume from 2012 (Nguyen et al., 2021). The port's proximity to industrial clusters and its integration with international shipping routes have been key to this growth. Vietnam's cruise industry has also flourished, with Halong Bay and Ho Chi Minh City collectively welcoming over 500,000 passengers annually before COVID-19 (McCoy, 2021). Recent investments in green logistics and automation position the port as a regional leader.

2.6 Sihanoukville, Cambodia

Sihanoukville Autonomous Port is Cambodia's primary deep-sea facility, handling 700,000 TEUs in 2022, up from 300,000 TEUs in 2012 (Ung et al., 2024). Under China's Belt and Road Initiative, the port has received substantial investments to modernise infrastructure and expand capacity (Han & Lim, 2021). Cruise tourism has also grown, with the port accommodating 50,000 passengers annually.

Sihanoukville's development is central to Cambodia's ambition of becoming a Southeast Asian trade hub.

2.7 Yangon, Myanmar

Yangon's Thilawa Port, Myanmar's principal maritime gateway, handled over 500,000 TEUs in 2022, doubling its capacity from a decade ago (Griffiths, 2023). Despite political and economic challenges, the port has focused on modernisation efforts, including expanded container yards and streamlined customs processes (Pushp & Ahmed, 2023). While cruise tourism remains limited, the port's regional ferry connections have supported increased passenger movements.

2.8 Regional Insights

These ports highlight the diversity and strategic importance of ASEAN's maritime infrastructure. While global leaders like Singapore continue to set benchmarks for efficiency, emerging players such as Sihanoukville and Thilawa are leveraging investments and regional initiatives to bolster their capabilities. ASEAN's ports have demonstrated significant cargo throughput and passenger traffic growth, reflecting the region's evolving role in global trade and tourism (Nguyen & Woo, 2022).

While ASEAN ports collectively contribute to the region's economic development and global trade networks, their importance and resilience vary significantly due to diverse factors. Ports like Singapore and Port Klang are critical transhipment and logistics hubs, leveraging advanced technologies and strong hinterland connectivity (Yap & Yang, 2024). Others such as Sihanoukville and Thilawa are emerging players that rely on targeted investments to enhance their capabilities. These differences in strategic importance reflect the varying roles ASEAN ports play within the global maritime industry.

In addition to their differing significance levels, ASEAN ports face various challenges. Climate change is a significant concern, with rising sea levels, extreme weather events, and shifting trade patterns posing risks to port infrastructure and operations (Yap & Yang, 2024). Low-lying facilities are particularly vulnerable to flooding, disrupting activities and leading to long-term damage. Economic factors such as global supply chain disruptions, fluctuating trade volumes and the relocation of manufacturing bases also influence port activity, particularly in export-heavy economies like Thailand and Vietnam.

Political and geopolitical risks further complicate the situation. Political instability, trade disputes and regional tensions can disrupt port operations and deter long-term investments, hindering the ability of some ports to achieve their full potential

(Hui, 2022). These varied impacts highlight the need to comprehensively understand how local vulnerabilities affect the broader global shipping network.

This diversity in importance and vulnerability underscores the necessity of a robust assessment framework, as introduced in Sect. 3. Such a framework is essential for evaluating the resilience of the global shipping network in the face of localised vulnerabilities, addressing the multifaceted challenges faced by ASEAN ports, and ensuring their sustained role in the global maritime ecosystem.

3 A Vulnerability Assessment Framework for a Global Shipping Network Experiencing Different Levels of Local Vulnerabilities

This section demonstrates the proposed methodology by analysing the ASEAN cruise port network, focusing on the influence of the COVID-19 pandemic (Poo et al., 2024a). The ASEAN region, known for its diverse cruise destinations and strategic maritime significance, offers an ideal case for evaluating network vulnerability and resilience (Poo & Yang, 2024). By incorporating the impacts of COVID-19, such as reduced ship capacities and disrupted itineraries, the analysis provides insights into the changes in port centrality and network efficiency. The methodology leverages real-world data from ASEAN cruise itineraries during the pandemic and recovery periods, offering a comprehensive view of how local vulnerabilities impact global cruise networks. The methodology has other modified versions such as comparing the new versions with added routes (Poo et al., 2024b), remodelling the potential changes in throughput and travel duration (Poo & Yang, 2024), and implying a risk assessment based on the frequency and severity on a trade route (Poo et al., 2024c).

3.1 Methodology

The vulnerability of cruise shipping networks can be analysed using centrality measures, highlighting different aspects of a port's importance within the network. Degree centrality identifies ports with the most direct connections, indicating their role in maintaining communication and tourism. Closeness centrality reflects a port's proximity to others, showing its efficiency in facilitating network operations. Betweenness centrality highlights ports acting as intermediaries on the shortest paths between others, showcasing their importance in maintaining network functionality. These measures alone have limitations, so this study uses a combination of centralities via the Borda Count method to comprehensively capture various aspects of port influence.

Additionally, the local port-level impacts of risks, such as the COVID-19 pandemic, are integrated into the network using the COVID-19 Performance Index (CPI). This approach adjusts link weights to reflect reduced ship capacity during crises.

The methodology involves four steps:

- Structuring the ASEAN cruise shipping network.
- Modelling the network using Social Network Analysis (SNA).
- Conducting vulnerability assessments without COVID-19 impacts.
- Reassessing vulnerability with COVID-19 impacts by incorporating the CPI.

The centrality measures, Borda Count, and local vulnerability adjustments form a robust framework to evaluate network resilience.

3.2 *Formulation*

The weighted SNA formulation, serving as the basis for network-level vulnerability assessment, is introduced to clarify the underlying mechanism. Drawing on graph theory, the network is conceptualised as a collection of ports (vertices or nodes) interconnected by links (edges or lines) (Biggs, 1993). This structure is represented by an $N \times N$ adjacency matrix (W), where rows and columns correspond to ports, and matrix elements capture the presence or weight of connections between them. Centrality metrics, including degree, closeness, and betweenness, are computed to evaluate each port's importance. These metrics are then integrated using the Borda Count method, yielding a composite score that reflects the overall influence and significance of each port within the network.

$$W = \begin{bmatrix} w_{11} & \cdots & w_{1N} \\ \vdots & \ddots & \vdots \\ w_{N1} & \cdots & w_{NN} \end{bmatrix} \tag{1}$$

The unnormalised weight w_{ij} is set to 0 if there is no connection between ports i and j. For undirected networks without loops, the adjacency matrix is symmetric ($w_{ij} = w_{ji} \geq 0$), and all diagonal elements are zero ($w_{ii} = 0$). Degree centrality quantifies the number of connections a port has in the network, with in-degree ($D(in)_i$) representing incoming links and out-degree ($D(out)_i$) representing outgoing links, expressed as Eq. (2) and (3), respectively, where N is the total number of ports. Closeness centrality measures the efficiency of a port in reaching others, with in-closeness ($C(in)_i$) and out-closeness ($C(out)_i$) defined as Eqs. (4) and (5). Betweenness centrality captures the extent to which a port lies on the shortest paths between others, calculated as B_i in Eq. (6) where $w(s, t| i)$ is the number of shortest paths between ports sss and ttt passing through port iii, and $w(s, t)$ is the total number of shortest paths between s and t.

$$D(in)_i = \sum_{h}^{N} w_{hi} \tag{2}$$

$$D(out)_i = \sum_{j}^{N} w_{ij} \tag{3}$$

$$C(in)_i = \frac{N-1}{D(in)_i} = \frac{N-1}{\sum_{h}^{N} w_{hi}} \tag{4}$$

$$C(out)_i = \frac{N-1}{D(out)_i} = \frac{N-1}{\sum_{j}^{N} w_{ij}} \tag{5}$$

$$B_i = \sum_{\substack{s,t \in N \\ s,t \neq i}} \frac{w(s,t|i)}{w(s,t)} \tag{6}$$

The Borda Count method integrates various centrality measures into a single ranking, offering simplicity and clarity in the calculation process. Higher values derived from Eqs. (2)–(6) correspond to higher rankings for cruise ports. The ranking positions denoted as $R_{D(in)}$, $R_{D(out)}$, $R_{C(in)}$, $R_{C(out)}$, R_B, represent the rank of each port based on degree (in and out), closeness (in and out), and betweenness centrality, respectively. These rankings are determined independently for each centrality measure. The count scores $S_{D(in)}(i)$, $S_{D(out)}(i)$, $S_{C(in)}(i)$, $S_{C(out)}(i)$, $S_B(i)$are calculated based on the ranking positions of each centrality for the i-th port, as detailed in Eqs. (6)–(10).

$$S_{D(in)}(i) = N - R_{D(in)}(i) + 1 \tag{7}$$

$$S_{D(out)}(i) = N - R_{D(out)}(i) + 1 \tag{8}$$

$$S_{C(in)}(i) = N - R_{C(in)}(i) + 1 \tag{9}$$

$$S_{C(out)}(i) = N - R_{C(out)}(i) + 1 \tag{10}$$

$$S_B(i) = N - R_B(i) + 1 \tag{11}$$

The degree and closeness centrality measure rankings account for both in and out directions. For ensuring a balanced contribution from degree, closeness, and betweenness centrality, each centrality type is assigned an equal weight of one-third in the final score calculation, as represented in Eq. (12).

$$S_O(i) = \frac{\frac{S_{D(in)}(i) + S_{D(out)}(i)}{2} + \frac{S_{C(in)}(i) + S_{C(out)}(i)}{2} + S_B(i)}{3} \tag{12}$$

The local port-level impact (*LI*) of the new link weight w'_{ij} is formulated in Eq. (13) and Eq. (14) to overview the COVID-19 impact by multiplying w_{hi} and $w_{ij}$with CPI (Delis et al., 2023) as LI of a cruise-calling port state to reflect the impact of COVID-19 on the reduced ship capacity, respectively.

$$w'_{hi} = w_{hi} \times LI / 100 \tag{13}$$

$$w'_{ij} = w_{ij} \times LI / 100 \tag{14}$$

The link weights are modified using the CPI to reflect capacity reductions during COVID-19. This adjustment enables comparative analyses of the network's performance under both normal and pandemic conditions.

3.3 *Structuring the ASEAN Cruising Network*

Cruise itineraries during two periods—September 2021 to August 2022 (COVID-19 period) and April 2023 to March 2024 (normal period)—were collected from CruiseTimetables. The data includes 243 ports, with 81 in ASEAN countries and 162 non-ASEAN ports connected through 201 itineraries. Annual link capacities were calculated by aggregating ship capacities. The network is bi-directional, with connections depending on the direction of travel.

3.4 *Modelling the ASEAN Cruise Shipping Network*

The network was modelled using UCINET 6 for visualisation and SNA. Centrality measures were computed for each port to assess their roles within the network. Adjusted link weights reflecting COVID-19 impacts were incorporated for comparative analysis. Full details can be found in Poo et al. (2024a).

4 Results

4.1 Vulnerability Assessment Without COVID-19 Impact

Using the Borda Count, Singapore ranks as the most influential port, followed by Phuket, Ho Chi Minh City, and Penang as shown in Table 1. Furthermore, the removal of top-ranked ports from the network confirmed their significant impact on global efficiency, validating the centrality rankings.

4.2 Vulnerability Assessment With COVID-19 Impact

The incorporation of CPI showed changes in port rankings, with ports in countries with lower CPI values, such as Indonesia, becoming less centralised. Conversely, ports in high-CPI countries like Vietnam and Singapore gained prominence. This highlights the varying impacts of local vulnerabilities on global network resilience.

The countries (i.e. the Philippines and Indonesia) with lower CPI are expected to become less centralised. Indonesian ports, namely Semarang, Komodo Island, and Surabaya, have experienced significant drops. However, it has also been noticed that Bali and Manila, two essential ports in the region, remain critical in the network

Table 1 Top 20 ports by multiple-centrality assessment

Port	Rank	Score
Singapore	1	240
Phuket, Thailand	2	233
Ho Chi Minh City, Vietnam	3	232
Penang, Malaysia	3	232
Klang (Kuala Lumpur), Malaysia	5	228
Bali, Indonesia	6	224
Ko Samui, Thailand	7	214
Laem Chabang (Bangkok), Thailand	8	208
Da Nang, Vietnam	9	206
Lombok, Indonesia	10	204
Semarang, Indonesia	11	200
Langkawi, Malaysia	12	198
Yangon, Myanmar	12	198
Bandar Seri Begawan, Brunei	14	196
Nha Trang, Vietnam	14	196
Manila, Philippines	16	195
Komodo Island, Indonesia	17	191
Sihanoukville, Cambodia	18	185
Surabaya, Indonesia	18	185
Pulau We, Indonesia	20	183

after the significant drop in weights. Langkawi, with less efficiency deduction, is raised significantly in rank. Besides, it has been noticed that ports with higher CPIs, such as Vietnam, Thailand, and Singapore, constantly increase the importance of the network. The changes among those countries with medium CPI are unpredictable. Langkawi from Malaysia has become more critical, but Klang has the opposite change.

Furthermore, Yangon and Bandar Seri Begawan have become more centralised, with less CPI than ports in Malaysia. The result represents that network-level vulnerability and local port-level vulnerability of the whole network are changing differently. The complete picture of the COVID-19 impacts on the global cruise shipping network can be obtained by integrating the methodology, which can be used in the scenarios assessing other or multiple impacts (Table 2).

Table 2 Top 20 ports about network-level vulnerability and port-level vulnerability with CPI

	Network-level vulnerability				Port-level vulnerability	
Port	Rank	Change	Score	Change	Rank	CPI
Singapore	1	0	240	0	3	73.3
Phuket, Thailand	2	0	233	0	2	82.6
Ho Chi Minh City, Vietnam	3	0	232	0	1	90.8
Penang, Malaysia	3	0	232	0	4	69.9
Klang (Kuala Lumpur), Malaysia	5	0	227	-1	4	69.9
Bali, Indonesia	6	0	222	−2	8	25.8
Ko Samui, Thailand	7	0	214	0	2	82.6
Langkawi, Malaysia	8	↑4	212	+14	4	69.9
Laem Chabang (Bangkok), Thailand	9	↓1	208	0	2	82.6
Da Nang, Vietnam	10	↓1	207	+1	1	90.8
Yangon, Myanmar	11	↑1	201	+3	5	62.4
Bandar Seri Begawan, Brunei	12	↑2	200	+4	6	61.1
Lombok, Indonesia	13	↓3	199	−5	8	25.8
Nha Trang, Vietnam	14	0	196	0	1	90.8
Semarang, Indonesia	15	↓4	194	−6	8	25.8
Manila, Philippines	16	0	193	−2	7	30.6
Sihanoukville, Cambodia	17	0	188	+3	6	61.1
Komodo Island, Indonesia	18	↓1	185	−6	8	25.8
Surabaya, Indonesia	19	↓1	177	−8	8	25.8
Kotakinabulu, Malaysia	19	↑2	177	0	4	69.9
Malacca, Malaysia	19	↑3	177	+4	4	69.9

Table 3 Top 20 ports changes between the COVID-19 period and the normal period

Port	COVID-19 period		Normal period			
	Rank	Score	Rank	Change	Score	Change
Singapore	1	240	1	0	240	0
Phuket, Thailand	2	233	7	↓5	216	−17
Ho Chi Minh City, Vietnam	3	232	2	↑1	230	−2
Penang, Malaysia	3	232	9	↓6	212	−20
Klang (Kuala Lumpur), Malaysia	5	228	3	↑2	229	1
Bali, Indonesia	6	224	4	↑2	228	4
Ko Samui, Thailand	7	214	16	↓9	198	−16
Laem Chabang (Bangkok), Thailand	8	208	13	↓5	204	−4
Da Nang, Vietnam	9	206	6	↑3	219	13
Lombok, Indonesia	10	204	11	↓1	211	7
Semarang, Indonesia	11	200	14	↓3	199	−1
Langkawi, Malaysia	12	198	9	↑3	212	14
Yangon, Myanmar	12	198	33	↓21	150	−48
Bandar Seri Begawan, Brunei	14	196	24	↓10	168	−28
Nha Trang, Vietnam	14	196	5	↑9	226	30
Manila, Philippines	16	195	12	↑4	208	13
Komodo Island, Indonesia	17	191	8	↑9	215	24
Sihanoukville, Cambodia	18	185	25	↓7	167	−18
Surabaya, Indonesia	18	185	20	↓2	176	−9
Pulau We, Indonesia	20	183	29	↓9	154	−29

4.3 *Comparative Analysis of COVID-19 and Normal Periods*

Post-pandemic, major cruise ports like Singapore and Bali have regained their prominence, while smaller ports such as Phuket and Penang have experienced a decline in centrality. This trend highlights the recovery favouring more prominent cruise destinations as demand surged after the pandemic. Table 3 presents a comparative analysis of cruise port networks during the COVID-19 period and normal period operations, showcasing the shifts caused by the pandemic. Southeast Asia marked a significant milestone with its first post-pandemic port call in July 2022. By 2023 and 2024, all ASEAN countries had re-established cruise port operations, reflecting a comprehensive regional recovery.

5 Discussion

Building maritime transport resilience in ASEAN offers numerous benefits that can significantly enhance the region's economic and strategic importance in the global supply chain. Resilient ports ensure uninterrupted trade flows, strengthening

ASEAN's position as a critical hub for global connectivity. Major ports like Singapore and Port Klang, known for their efficiency and advanced infrastructure, provide stability to supply chain operations, attracting investment and fostering industrial growth. Additionally, implementing resilience measures, such as climate-adaptive designs and green logistics solutions, enables ports to mitigate the impacts of climate change, reducing vulnerabilities to extreme weather events and rising sea levels. Collaborative regional efforts to enhance resilience further promote knowledge sharing and resource pooling, enabling the ASEAN countries to tackle shared challenges collectively and efficiently.

Despite these advantages, building resilience in maritime transport also presents significant challenges. High initial investment costs for upgrading infrastructure and implementing advanced technologies can strain financial resources, particularly for smaller or emerging ports like Sihanoukville and Thilawa. This financial disparity risks widening the development gap between well-established hubs and less developed ports, creating an uneven distribution of economic benefits across the region. Moreover, the shift towards digitalisation and automation may face resistance due to a lack of technical expertise, cybersecurity risks, and workforce retraining needs, posing operational hurdles.

Geopolitical tensions further complicate efforts to build resilience. Disputes over maritime boundaries, trade policies and regional power dynamics may hinder cooperative initiatives, impacting long-term progress. Environmental concerns also pose a dilemma, as increased port activity and expansion projects, if not managed sustainably, could lead to habitat destruction and pollution, undermining the very goals of building climate-adaptive and sustainable infrastructure.

Nevertheless, ASEAN's maritime resilience is immensely important to the global supply chain. The region is a critical crossroads for trade routes connecting Asia, Europe and the Americas. Ports like Singapore and Ho Chi Minh City act as vital nodes, facilitating the efficient movement of goods and raw materials worldwide. Disruptions in ASEAN's maritime transport systems can trigger ripple effects across global supply chains, delaying shipments, increasing costs and destabilising industries. Strengthening resilience in this region is not just a regional priority but a global imperative, ensuring the stability and sustainability of interconnected economies.

6 Conclusion

Building resilience in ASEAN's maritime transport and supply chain networks is both a necessity and an opportunity. The advantages, including enhanced connectivity, economic stability and climate adaptation, underscore the critical role of resilience in securing regional and global trade. However, challenges such as high costs, unequal development and geopolitical risks must be addressed through comprehensive policies and regional cooperation.

Given ASEAN's strategic importance in the global supply chain, ensuring the resilience of its maritime infrastructure is vital for maintaining trade efficiency and economic stability worldwide. By investing in climate-adaptive infrastructure, embracing digitalisation and fostering regional collaboration, ASEAN can position itself as a model for sustainable and resilient maritime systems. This development will not only bolster the region's economic competitiveness but also contribute to the stability and growth of the global economy, ensuring that maritime networks remain robust in the face of future uncertainties.

References

Ayuttacorn, A. (2023). Chinese investor networks and the politics of infrastructure projects in the Eastern Economic Corridor in Thailand. *Advances In Southeast Asian Studies, 16*, 233–250.

Biggs, N. (1993). *Algebraic graph theory*. Cambridge University Press.

Carruthers, R., Bajpai, J. N., & Hummels, D. (2003). Trade and logistics: An East Asian perspective. In *East Asia integrates: A trade policy Agenda for shared growth* (pp. 117–137).

Delis, M. D., Iosifidi, M., & Tasiou, M. (2023). Efficiency of government policy during the Covid-19 pandemic. *Annals of Operations Research, 328*, 1287–1312.

Griffiths, M. (2023). Citizens and land: Socioeconomic effects of relocation and resettlement by the Thilawa special economic zone. *Journal Of Burma Studies, 27*, 291–328.

Han, S. S., & Lim, Y. (2021). China's belt and road initiative (Bri) In Cambodia. In *International perspectives on the belt and road initiative*. Routledge.

Handani, N. D., Williady, A., & Kim, H.-S. (2022). An analysis of customer textual reviews and satisfaction at luxury hotels in Singapore's Marina Bay Area (Sg-Clean-Certified Hotels). *Sustainability, 14*, 9382.

Hui, H. C. (2022). The long-run effects of geopolitical risk on foreign exchange markets: Evidence from some Asean Countries. *International Journal of Emerging Markets, 17*, 1543–1564.

Idris, H. (2022). Southeast asian port development: Policy and initiatives towards achieving 2030 Agenda on sustainable development goals. *Akademika, 92*, 129–142.

Lee, Y. (2018). Maritime merchandise trade In Southeast Asia: Opportunities and challenges. *Maritime Affairs: Journal of the National Maritime Foundation of India, 14*, 51–69.

Mccoy, E. R. (2021). *A war tour of Viet Nam: A cultural history*. Mcfarland.

Mohd Rozar, N., Sidik, M. H., Razik, M. A., Ahmad Kamaruddin, S., Rozar, M. K. A. M., Usman, I., & Alown, B. E. (2023). A hierarchical cluster analysis of port performance in Malaysia. *Maritime Business Review, 8*, 194–208.

Nguyen, P. N., & Woo, S.-H. (2022). Port connectivity and competition among container ports in Southeast Asia based on social network analysis and topsis. *Maritime Policy & Management, 49*, 779–796.

Nguyen, T. L. H., Park, S.-H., Kim, Y., & Yeo, G.-T. (2021). An efficiency analysis of container terminals in Southern Vietnam using dea dynamic efficiency evaluation. *The Asian Journal Of Shipping And Logistics, 37*, 329–336.

Panahi, R., Gargari, N. S., Lau, Y.-Y., & Ng, A. K. (2022). Developing A resilience assessment model for critical infrastructures: The case of Port in tackling the impacts posed by the Covid-19 pandemic. *Ocean & Coastal Management, 226*, 106240.

Poo, M. C.-P., & Yang, Z. (2024). Optimising the resilience of shipping networks to climate vulnerability. *Maritime Policy & Management, 51*, 15–34.

Poo, M. C.-P., Yang, Z., Dimitriu, D., Qu, Z., Jin, Z., & Feng, X. (2021). Climate change risk indicators (Ccri) for seaports in the United Kingdom. *Ocean & Coastal Management, 205*, 105580.

Poo, M. C.-P., Yang, Z., & Lau, Y.-Y. (2024a). Vulnerability analysis of cruise shipping in Asean Countries facing Covid-19 pandemic. *Ocean & Coastal Management, 248*, 106919.

Poo, M. C.-P., Yang, Z., Lau, Y.-Y., & Jarumaneeroj, P. (2024b). Assessing the impact of arctic shipping routes on the global container shipping network's connectivity. *Polar Geography, 47*, 219–239.

Poo, M. C.-P., Zhang, W., Kamalian, L., Wang, T., Lau, Y.-Y., & Xu, T. Z. (2024c). Resilience of Chinese Ports to tropical cyclones: Operational efficiency and strategic importance. *Climate, 12*, 214.

Pushp, P., & Ahmed, F. (2023). The Global value chain: Challenges faced by Asean least developed countries. *Journal of Policy Modeling, 45*, 1223–1245.

Rachman, A., & Wahid, A. (2023). The Influence of service quality, relationship quality on Port performance which has implications for customer satisfaction at the Port Tg. Priok Jakarta Indonesia. *International Journal Of Advanced Multidisciplinary, 2*, 475–487.

Rozman, G. (2022). *Strategic triangles reshaping international relations in East Asia*. Taylor & Francis.

Sukhotu, V. (2021). Trade promotion from Thailand To China as a result of a new train route. *Journal Of Business & Economics Review (Jber), 6*, 1.

Ung, H., Rim, S., & Po, S. (2024). *China's involvement In Cambodian ports and its impacts with special attention to Dara Sakor. Chinese Overseas Ports In Southeast And South Asia*. Routledge.

Verhaeghe, R., Halim, R. A., & Tavasszy, L. (2021). Optimizing the efficiency of the future maritime transport network of Indonesia. In *Freight transport modeling in emerging countries*. Elsevier.

Wan, C., Zhao, Y., Zhang, D., & Yip, T. L. (2021). Identifying important ports in maritime container shipping networks along the Maritime silk road. *Ocean & Coastal Management, 211*, 105738.

Yang, Z., Ng, A. K., Lee, P. T.-W., Wang, T., Qu, Z., Rodrigues, V. S., Pettit, S., Harris, I., Zhang, D., & Lau, Y.-Y. (2018). Risk and cost evaluation of port adaptation measures to climate change impacts. *Transportation Research Part D: Transport and Environment, 61*, 444–458.

Yap, W. Y., & Yang, D. (2024). Hub Port choice and shipping connectivity In Southeast Asia during Covid-19 pandemic: Implications for post-pandemic competition landscape. *Maritime Policy & Management, 51*, 1334–1349.

Yeo, I. 2024. Container port selection in the Asean Region: Korean Shipping Companies' perspective.

Sustainable Development of Blockchain Technology for Vaccine Supply Chain

1 Introduction

Inspired by the insight of the potentiality of preventive measures against the widespread COVID-19 pandemic since January 2020 and the enhancement in increasing the logistics firms' and healthcare specialists' ability to encounter vaccine vulnerabilities that cause global health risks (e.g. inferior quality of vaccine due to inappropriate storage, location, vibration, and temperature), a cloud-based blockchain technology has been proposed and designed for building resilient vaccine supply chain. The studies of blockchain technology in the supply chain were first reported in 2017 (Paliwal et al., 2020). Since then, the growing attention to blockchain technology in the supply chain has laid the foundations for studies of blockchain features (e.g. security, decentralization, shareability, immutability, trust, traceability, and transparency) to improve supply chain management.

The basic structure of a blockchain contains a block and a chain. Initially, when a stakeholder creates a transaction such as a shipment schedule and administrative process (e.g. issue of purchase orders, bills of lading, invoices, packing lists, certificates of origin, insurance certificates, and declarations of dangerous products), that transaction is recorded in a block. This first block keeps storing new transaction records until this first block's storage capacity is reached. Then, the second block is created and connected to the first block. This second block keeps recording new transactions until its storage capacity is met. Then, the third block is created and linked to the second block. As more and more new blocks are created and appended to an existing chain of blocks, a sequential chain of blocks is formed. This sequential chain of blocks is a blockchain. Wong et al. (2021) identified the following major blockchain features that help to improve supply chain operations:

- Security—Each block in the blockchain consists of two parts: a header and a body. The header contains an identity of the block which is determined by a hash of its previous block, a timestamp and a Merkle root. The hash can represent the

Y.-y. Lau et al., *Maritime Transport and Supply Chain Resilience*,
https://doi.org/10.1007/978-3-032-07566-6_7

cryptographic signature which can be used to validate transactions stored in the body. The Merkle root is built by a cryptography method based on a Merkle tree (Merkle, 1988) which is a data structure that can verify the block body transactions efficiently (Azzi et al., 2019). As an example, the security feature of the blockchain application developed by IBM Food Trust, JD.com, Tsinghua University, and Wal-Mart ensures food safety and security in its food supply chains in China (Park & Li, 2021).

- Decentralization and shareability—When creating a transaction, that transaction is recorded in the body of a block and replicated. This replica is then distributed, shared and agreed among the stakeholders involved in the transaction in a peer-to-peer network. In this peer-to-peer network, all the involved stakeholders are peers having the same privilege. There is no centralized control of distributing and sharing transactional activities in that network. As an example, the decentralization and shareability features of the TradeLens blockchain used by Maersk allow the exchange of documents (e.g. bills of lading, letters of credit, and forwarders' certificates of transport) among a large variety of supply chain stakeholders (e.g. banks, suppliers, port authorities and buyers) (Wong et al., 2021).
- Immutability and trust—Immutability refers to the situation in which a validated transaction record cannot be updated or changed in any way. If modification or correction on the validated transaction record is necessary, a new transaction that issues the modification or correction is created, recorded, and validated in the blockchain peer-to-peer network. Each block timestamps transaction records store the records securely by public-key cryptography (i.e. hash) and validate them. In this way, the transaction records can be verified and trusted by the stakeholders involved in the transactions in the blockchain P2P network (Varriale et al., 2020). For example, Energy Blockchain Labs in China has developed a blockchain platform that helps organizations in the supply chain to build trust among them by monitoring carbon footprints and meeting carbon emission reduction quotas (Saberi et al., 2019).
- Traceability and transparency—The stakeholders involved in the transactions can duplicate and distribute their transaction records in a secure and immutable way as their transaction records are hashed, timestamped and stored in a blockchain and confirmed by the consensus mechanism in the blockchain peer-to-peer network (Feng et al., 2020), imposing impossibility for reversing the blockchain's entire historical transaction records under the scrutiny of all the stakeholders who can access the distributed and shared transaction records (Tapscott & Tapscott, 2016), and providing traceability (Park & Li, 2021), audibility and transparency (Yeung et al., 2021) of keeping transaction records in the blockchain peer-to-peer network. For example, the traceability and transparency features of SAP's blockchain hub help to reduce the cost of tracing medical products in supply chains (O'Donnell, 2019).

It is expected that if more and more rapidly expanding blockchains emerge in the blockchain peer-to-peer network, then the stakeholders have to face the challenge of the technical sustainability of the blockchain technology in terms of its availability, scalability, latency, throughput, and big data storage capacity (Zhao et al., 2019). To tackle this challenge, Wong et al. (2021) designed a cloud-based blockchain architecture that applies to vaccine supply chain management. Meanwhile, tremendous processes and data derived from a variety of stakeholders in the blockchain P2P network provide sources for big data analytics which can be enhanced by machine learning. To take the opportunity of big data analytics when adopting blockchain technology, Wong et al. (2021) proposed integrating machine learning into the cloud-based blockchain architecture.

For technical sustainability, as blockchain applications are still in their infancy stage, users' hesitation to adopt the blockchain technology is expected. If users do not accept the adoption of blockchain technology, that technology cannot be sustained, resulting in a technical sustainability challenge. Knowing the technical sustainability issues, the researchers are concerned about the sustainable development of such a new cloud-based blockchain technology for vaccine supply chain applications for the well-being of a community. The community's well-being is the public health in the Greater Bay Area through the aid of the vaccine supply chain as the Greater Bay Area development is in accord with the government's strategic plan for innovation-driven development and commitment as an international area ideal for living, working, and traveling.

In general, this book chapter is split into three main sections. Section 1 gives an overview of the research background and settings. Then, Sect. 2 conducts a literature review including technical sustainability and the development of vaccines. After that, Sect. 3 provides the discussion and conclusion.

2 Literature Review

The researchers first set the following inclusion criteria and appropriate search terms, and used them to search through the Internet search engines (e.g. Google, ProQuest, Scopus, and Web of Science) in order to extract the relevant studies in literature:

For cloud-based blockchain for vaccine supply chain applications in the Greater Bay Area,

- Studies related to Task-Technology Fit and the Unified Theory of Acceptance and Use of Technology and technical sustainability.

The search results show that there is no previous relevant research. Then, fewer and less stringent terms by taking out some keywords were tried. Until the terms "blockchain for supply chain" were used for the above search, some relevant previous studies were found.

For the exploration of technical sustainability, the keywords "technology adoption/acceptance", "task-technology fit/TTF, and "unified theory of acceptance and use of technology/UTAUT" were added to the search. Then, eight relevant previous studies related to the technical sustainability exploration in this research study are found. These are the studies by Alazab et al. (2021), Francisco and Swanson (2018), Lee et al. (2019), Queiroz et al. (2021), Queiroz and Wamba (2019), Wahab et al. (2020), Wamba and Queiroz (2019) and Wong et al. (2020). Different from these studies that applied either UTAUT or its revised models, the proposed study combines TTF and UTAUT into a new TTF-UTAUT model which is more appropriate as whether the technology fits the users' task performance is also considered (Fig. 1).

2.1 Technical Sustainability

Many previous studies (e.g. Park & Li, 2021; Varriale et al., 2020) focused on how blockchain features benefit supply chain operations. However, if the supply chain stakeholders (e.g. supply chain operations personnel) do not accept utilizing blockchain technology, that technology cannot help to improve the supply chain operations. The mere utilization of the technology cannot guarantee better performance. For example, students cannot learn technical drawing skills with the use of a word processing app as its capabilities do not fit the technical drawing tasks and performance. Therefore, whether the technology functionality fits performance is also important. As blockchain technology is introduced for the vaccine supply chain in this book chapter, whether that technology fits the vaccine supply chain

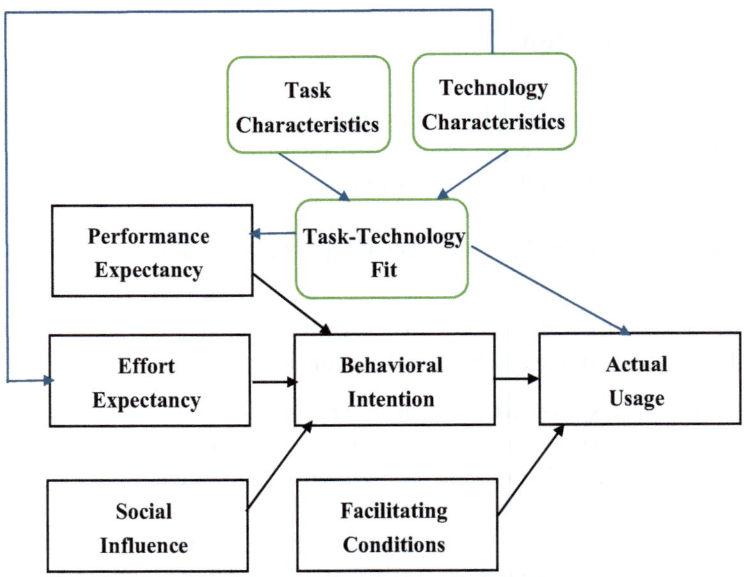

Fig. 1 Integration of Task-Technology Fit and the Unified Theory of Acceptance and Use of Technology for Technical Sustainability Exploration

stakeholders' task performance and whether the stakeholders accept using that technology should be investigated.

For the technical sustainability exploration, the researchers consider Goodhue and Thompson's (1995) Task-Technology Fit theoretical construct which asserts the influence of a good fit of the technology on the tasks (e.g. supply chain operations) to be performed. In this theory, a construct called task characteristics includes the functions of a task. Another construct is called technology characteristics which refers to the functions and features of a technology. The degree of a match between task characteristics and technology characteristics is described by task-technology fit. Both task characteristics and technology characteristics determine Task-Technology Fit, which in turn determines the technology users' performance and utilization of a technology.

Also, the researchers consider Venkatesh et al.'s (2003) Unified Theory of Acceptance and Use of Technology. The Unified Theory of Acceptance and Use of Technology contains moderating (or indirect) effects (i.e. gender, age, experience, and voluntariness of use), but they are not examined in this book chapter as the researchers intend to obtain the findings in this study that can apply to any gender and any age, and expect there is not much difference in the experience of using this new cloud-based blockchain technology and the voluntariness of use as the users are supposed to use the technology which has been adopted in their organizations. As theorized by the Unified Theory of Acceptance and Use of Technology, the users' acceptance of the cloud-based blockchain technology is indicated by their actual usage behaviour of that technology which is determined by their behavioural intention to use that technology and facilitating conditions. The users' behavioural intention is in turn determined by their own three perceptions: (1) performance expectancy, which is the degree of the users' belief that using the cloud-based blockchain technology can enhance their task performance; (2) effort expectancy, which is the degree of the users' perception of the digital literacy, self-efficacy and ease of use of the cloud-based blockchain system; and (3) social influence, which is the extent to which the users perceive that the people around them, such as supply chain parties, business partners, supervisors, colleagues, friends and relatives, expect that they should perform the technology usage behaviour.

2.2 *The Development of Vaccines*

Vaccine development has undergone three main stages. In the first stage, we mainly focus on exploration, which adopts various research calculation models to affect the elements and antigen body of the vaccine. In the second stage, we concentrate on preclinical research, which employs cultured animals and cells to conduct experiments to assess the immunity and safety of the vaccine. After the indexes are supported on animals, human assessments will be carried out, and the third stage will be a large group test accordingly (Sharma et al., 2020). The five significant processes of the third stage are summarized in Table 1.

Table 1 The five main processes of vaccine development during the third stage

Stage	Key notes
1. Safety	Check the dose and test the vaccine on people with important resilience, safety, and unsusceptible response
2. Enlarge the safety	Vaccinate people of various age groups, and constantly re-test the dose and unsusceptible response, generally to make sure the unsusceptible efficacy and safety influence the dosage of the test for the next stage
3. Efficacy	Give vaccines to an abundance of people to assess efficacy
4. Review and approval	The outcomes of clinical trials require to be reviewed, and it is necessary to determine whether the vaccine is appropriate for further approval
5. Post-marketing monitoring	Tracking the potential negative responses after public use

2.3 *Vaccine in COVID-19*

As of mid-2020, the worldwide SARS-CoV-2 vaccine contained 158 vaccines, of which 135 were in the development initial or preclinical stage (Kaur & Gupta, 2020). The Food and Drug Administration has further confirmed the adoption of BioNTech/Pfizer and Moderna COVID-19 vaccines in immunization programmes. BioNTech/Pfizer employed over 43,000 participants, and Moderna employed 30,000 participants during the third clinical trial. The outcomes of the experiment indicate that the vaccine may decrease the COVID-19 infection rate and preserve recipients via giving immunity and creating antibodies against the COVID-19 virus (Mohamed et al., 2021). The United Kingdom was the first country to conduct an immunization program. Pfizer and Moderna adopt an innovative and novel technology with mRNA performing as the vaccine and the active substance includes RNA which is for cells to create protein as an antigen (Kumar & Gupta, 2020). Other vaccines adopt various kinds of antigens like attenuated viruses, inactivated viruses, and viral vectors. Also, Sinovac is developing an aluminium and inactivated adjuvant vaccine (Sharma et al., 2020).

The first COVID-19 vaccine was jointly developed by BioNTech and Pfizer and was confirmed in December 2020 (Saleh et al., 2021). The evolution of the COVID-19 vaccine is identified as one of the rapid developments in the history of vaccine science. People are anxious about the vaccine as it has failed to effectively immunize against viruses because of its speedy development. It may induce people to lack confidence in the vaccine. Thus, several people in different countries or cities have not been vaccinated.

3 Discussion and Conclusion

For a community to sustain itself, practices leading to technical performances must be implemented. This proposed study explores these practices in terms of the technical sustainability of implementing cloud-based blockchains for vaccine supply chain applications in the Greater Bay Area.

This proposed study is significant in the sense that a new and more appropriate Task-Technology Fit and the Unified Theory of Acceptance and Use of Technology model is adopted to explore the technical sustainability issue. At the same time, the literature does not address sufficient evidence of the applicability of the technical sustainable practices to the implementation of the cloud-based blockchain for vaccine supply chain applications in the Greater Bay Area. This proposed study has short-term, medium-term, and long-term impacts which are presented as follows.

For the short term (i.e. 1 to 3 years) in the future, this proposed study investigates the technical sustainability in terms of the user's intention to adopt the cloud-based blockchain for vaccine supply chain applications in the Greater Bay Area at this initial stage of implementing that technology. Whether this technology can be sustained depends on the technology users' intention to utilize the technology and their actual usage. Also, the idea of implementing the cloud-based blockchain technology for vaccine supply chain applications is still underway in the Greater Bay Area though there is a growing attention to adopting this new blockchain technology for supply chains around the world. In this regard, the concepts of cloud-based blockchain technology and its applicability in the vaccine supply chain in the Greater Bay Area can only be conveyed to the stakeholders. And so, the potential effect of the cloud-based blockchain for vaccine supply chain applications on sustainability performances can be analysed in this proposed study. The study is based on the participating stakeholders' conceptual perceptions instead of their actual experiences. However, their views and concerns are imperative for formulating the guidelines and blueprints for planning and designing the implementation of the cloud-based blockchain for vaccine supply chain applications in the Greater Bay Area.

For the medium term (i.e. 4–10 years) future, this proposed study attempts to cause the start-up blockchain use cases in the vaccine supply chain in the Greater Bay Area, and its findings are expected to provide new knowledge that facilitates the development of appropriate guidelines as well as orientation activities such as technology training for implementing the blockchain in the vaccine supply chain in the Greater Bay Area. With the actual development of the cloud-based blockchain technology for vaccine supply chain applications in the Greater Bay Area, the stakeholders' experiences of using that technology can be investigated to form technically sustainable practices. Moreover, more and more blockchain applications in other supply chains, for instance, food and drugs, to name but a few, in some other regions will be developed as time goes by. The proposed study can suggest directions for future development of the blockchain in other supply chains in these other regions. Furthermore, these findings provide teaching and learning curricula about the implementation of blockchain applications in supply chain management.

For the long term (i.e. over 10 years) in the future, a longitudinal study on technically sustainable practices can be carried out as the immutable transaction records stored in the blockchain provide reliable sources for that study. Also, the proposed research study can be extended to study sustainability performances when new technology is incorporated into blockchain technology. For example, tremendously immutable and distributed transaction records derived from blockchain applications provide sources for big data analytics. Future innovative blockchain applications with sophisticated machine learning algorithms can be developed to perform data analytics on such big data.

References

Alazab, M., Alhyari, S., Awajan, A., & Abdallah, A. (2021). Blockchain technology in supply chain management: An empirical study of the factors affecting user adoption/acceptance. *Cluster Computing, 24*(1), 83. https://doi.org/10.1007/s10586-020-03200-4

Azzi, R., Chamoun, R. K., & Sokhn, M. (2019). The power of a blockchain-based supply chain. *Computers and Industrial Engineering, 135*, 582–592.

Feng, J., Zhao, X., Chen, K., Zhao, F., & Zhang, G. (2020). Towards random-honest miners selection and multi-blocks creation: Proof-of-negotiation consensus mechanism in blockchain networks. *Future Generation Computer Systems—the International Journal of Science, 105*, 248–258.

Francisco, K., & Swanson, R. D. (2018). The supply chain has no clothes: Technology adoption of blockchain for supply chain transparency. *Logistics., 2*, 2. https://doi.org/10.3390/logistics2010002

Goodhue, D. L., & Thompson, R. L. (1995). Task technology fit and individual performance. *MIS Quarterly, 19*(2), 213–236.

Kaur, S. P., & Gupta, V. (2020). COVID-19 vaccine: A comprehensive status report. *Virus Research, 288*, 198114.

Kumar, G., & Gupta, S. (2020). Assessment of cold chain equipments and their management in government health facilities in a district of Delhi: A cross-sectional descriptive study. *Indian Journal of Public Health, 64*(1), 22.

Lee, C. C., Kriscenski, J. C., & Lim, H. S. (2019). An empirical study of behavioral intention to use blockchain technology. *Journal of International Business Disciplines, 14*(1), 1–21.

Merkle, R. C. (1988). A digital signature based on a conventional encryption function. In C. Pomerance (Ed.), *Advances in cryptology—CRYPTO '87, lecture notes in computer science* (pp. 369–378). Springer.

Mohamed, N. A., Solehan, H. M., Mohd Rani, M. D., Ithnin, M., & Che Isahak, C. I. (2021). Knowledge, acceptance and perception on COVID-19 vaccine among Malaysians: A web-based survey. *PLoS One, 16*(8), e0256110.

O'Donnell, J. (2019). *SAP blockchain service helps fight counterfeit drugs.* Retrieved October 22, 2021, from https://www.techtarget.com/searchsap/feature/SAP-blockchain-service-helps-fight-counterfeit-drugs

Paliwal, V., Chandra, S., & Sharma, S. (2020). Blockchain technology for sustainable supply chain management: A systematic literature review and a classification framework. *Sustainability, 12*(18), 7638.

Park, A., & Li, H. (2021). The effect of blockchain technology on supply chain sustainability performances. *Sustainability, 13*(4), 1726.

Queiroz, M. M., & Wamba, S. F. (2019). Blockchain adoption challenges in supply chain: An empirical investigation of the main drivers in India and the USA. *International Journal of Information Management, 46*, 70–82.

Queiroz, M. M., Wamba, S. F., De Bourmont, M., & Telles, R. (2021). Blockchain adoption in operations and supply chain management: Empirical evidence from an emerging economy. *International Journal of Production Research, 59*(20), 6087–6103.

Saberi, S., Kouhizadeh, M., Sarkis, J., & Shen, L. (2019). Blockchain technology and its relationships to sustainable supply chain management. *International Journal of Production Research, 57*(7), 2117–2135.

Saleh, A., Qamar, S., Tekin, A., Singh, R., & Kashyap, R. (2021). Vaccine development throughout history. *Cureus, 13*, e16635. https://doi.org/10.7759/cureus.16635

Sharma, O., Sultan, A. A., Ding, H., & Triggle, C. R. (2020). A review of the progress and challenges of developing a vaccine for COVID-19. *Frontiers in Immunology, 11*, 585354.

Tapscott, D., & Tapscott, A. (2016). *Blockchain revolution: How the technology behind bitcoin is changing money, Business and the World*. Penguin Random House.

Varriale, V., Cammarano, A., Michelino, F., & Caputo, M. (2020). The unknown potential of blockchain for sustainable supply chains. *Sustainability, 12*(22), 9400.

Venkatesh, V., Morris, M. G., Davis, G. B., & Davis, F. D. (2003). User acceptance of information technology: Toward a unified view. *MIS Quarterly, 27*(3), 425–278.

Wahab, S., Loo, Y., & Say, C. (2020). Antecedents of blockchain technology application among Malaysian warehouse industry. *International Journal of Logistics Systems and Management, 1*(1), 1. https://doi.org/10.1504/IJLSM.2020.10032312

Wamba, S. F., & Queiroz, M. M. (2019). The role of social influence in blockchain adoption: The brazilian supply chain case. *IFAC-PapersOnLine, 52*, 1715–1720.

Wong, L.-W., Tan, G. W.-H., Lee, V.-H., Ooi, K.-B., & Sohal, A. (2020). Unearthing the determinants of blockchain adoption in supply chain management. *International Journal of Production Research, 58*(7), 2100–2123.

Wong, S., Yeung, J., Lau, Y. Y., & So, J. (2021). Technical sustainability of cloud-based blockchain integrated with machine learning for supply chain management. *Sustainability, 13*, 8270.

Yeung, J., Wong, S., & Chan, H. H. (2021). Cloud infrastructure for public health blockchain. In B. Y. F. Fong & M. C. S. Wong (Eds.), *Routledge handbook of public health and the community* (pp. 254–263). Routledge.

Zhao, G., Liu, S., Lopez, C., Lu, H., Elgueta, S., Chen, H., & Boshkoska, B. M. (2019). Blockchain technology in Agri-food value chain management: A synthesis of applications, challenges and future research directions. *Computers in Industry, 109*, 83–99.

The Complexity of Adaptation Strategies for Maritime Transport

1 Introduction

Maritime transport is indispensable in connecting global supply chains, yet its infrastructure and operations face multifaceted threats. Climate change, in particular, presents significant risks, including rising sea levels, extreme weather events, and shifts in trade patterns. These challenges are compounded by global disruptions such as the COVID-19 pandemic and geopolitical conflicts, which strain logistics and demand rapid responses. This chapter analyses these threats and discusses how adaptation measures can mitigate them. The importance of stakeholder collaboration at various scales—from local port authorities to international organisations—is effectively addressed in response to these challenges.

2 Climate Risks to Maritime Transport Systems

Maritime transport systems are increasingly vulnerable to a wide range of climate threats, with evidence from multiple studies and reports underlining the scope of these challenges. Rising sea levels, documented by the IPCC and national climate agencies, threaten the structural integrity of ports and coastal zones (Mastrandrea et al., 2010). Extreme weather events such as hurricanes and typhoons, highlighted in reports by the World Meteorological Organization (WMO), disrupt port operations and damage vital infrastructure. Additionally, research from McIntosh and Becker (2019) mentions increased precipitation and flooding as critical risks, impairing access to ports and transportation networks. As Chhetri et al. (2016) reported, higher temperatures decrease key electronic systems' reliability and operational efficiency. Finally, snowing also affects the port operations as workers may be unable to return for their duty. Apart from that, any climate threat can affect the

Y.-y. Lau et al., *Maritime Transport and Supply Chain Resilience*,
https://doi.org/10.1007/978-3-032-07566-6_8

demand and supply of logistics from different regions and the shipping schedules (Song, 2021). Table 1, based on data from the IPCC, illustrates the climate threats faced by different regions and interconnected systems:

As critical global and regional trade nodes, UK ports face several key risks as listed in the climate change adaptation reporting requested by the Department for Environment, Food & Rural Affairs (DEFRA) under the Climate Change 2008 Act (Poo et al., 2021). The first-round reports inviting harbour authorities were released by DEFRA in 2011, the second-round reports were released in 2015 and 2016, and the third-round reports were released in 2022 and 2023.

UK ports are structured into key subsystems, each facing unique climate threats and requiring tailored adaptation strategies. The port infrastructure, encompassing terminals, berths, breakwaters, and operational systems, is at risk from high waves, storm surges, rising sea levels, and coastal erosion. Measures such as building or enhancing breakwaters, elevating port infrastructure, and installing protective flood defence systems are essential to mitigate these threats.

Transport connectivity, which includes the road and rail networks linking ports, is particularly vulnerable to flooding, landslides, and extreme weather events. Adaptation measures in this category include improving drainage systems, adopting thermal-resistant materials for infrastructure, and strengthening the foundations of bridges and culverts to withstand erosion and flooding.

Urban systems and communities near ports also face significant risks, such as increased heat, flooding, and air pollution. These challenges call for nature-based solutions, such as urban greening initiatives and beach nourishment programs, to mitigate erosion and heat risks.

Technological upgrades further enhance resilience across these subsystems. Enhanced cooling systems for temperature-sensitive equipment and flood monitoring systems linked to agency forecasts provide real-time risk management. Additionally, management improvements, such as expanding dredging programs, relocating vulnerable facilities, and diversifying land connections, ensure the operational continuity of UK ports.

These comprehensive adaptation measures highlight the importance of integrating infrastructure strengthening, climate-specific designs, nature-based solutions, and technological innovations to address the multifaceted threats climate change poses to UK ports.

3 Adaptation Strategies for Connected Systems

Rail networks are vital for port connectivity but are highly susceptible to extreme weather events (Palin et al., 2021). Upgrading track drainage systems is essential to handle the increasing volume and intensity of precipitation, reducing the risk of flooding and operational disruptions (Chen et al., 2021). Thermal expansion measures, such as adjusting the design standards for track materials, can prevent track buckling during extreme temperature events (Ngamkhanong, 2021). Additionally,

Table 1 Climate threats to regions and interconnected systems

Region/ System	Warming trend	Extreme temperature	Precipitation	Extreme precipitation	Damaging cyclones	Drying trend	Flooding	Snow cover	Sea-level rise	Storm surge	Ocean acidification	Sea surface temperature
Europe	✓	✓		✓					✓			
Asia	✓	✓		✓	✓	✓			✓	✓		
Australasia	✓	✓		✓	✓	✓			✓	✓		
North America	✓	✓	✓	✓	✓	✓			✓			
Central and South America	✓	✓	✓	✓			✓		✓	✓		✓
Polar regions	✓								✓			
Small islands	✓	✓		✓		✓			✓	✓	✓	
Transportation systems		✓		✓			✓	✓				
Coastal zone systems									✓	✓		

introducing flood risk monitoring systems for bridge foundations ensures timely interventions, minimising the risk of structural failures during extreme floods (Sasidharan et al., 2022).

Road networks are critical in maintaining access to ports and are particularly vulnerable to climate-induced challenges (Mikaelsson & Dzebo, 2023). Revising drainage systems to cope with heavy rainfall can mitigate the risks of waterlogging and unsafe driving conditions (Zhou et al., 2013). Enhancing material selection and updating construction standards will ensure resilience against extreme temperatures and storms (Rincon et al., 2024). Strengthening the foundations of bridges and culverts is equally important to protect against erosion caused by rising sea levels and heavy rainfall, preserving the integrity of essential transport infrastructure (Schulz et al., 2017).

Urban systems and communities surrounding ports are at the forefront of climate impacts, facing heat, flooding, and pollution risks (Becker et al., 2018). Heat risk mitigation strategies include introducing urban greening initiatives, green roofs, and enhancing public health access to shield vulnerable populations from extreme heat events (Sanchez and Reames, 2019). Early warning systems and nature-based solutions, such as sustainable urban drainage, are pivotal in reducing damage and enhancing resilience for flood risk management (Young et al., 2019). Finally, air and water quality improvements require stricter emissions regulations and adaptive planning for wastewater management, safeguarding both public health and urban ecosystems (Hughes et al., 2021).

4 The Need for Multi-stakeholder Collaboration

The private sector contributes significantly by developing green technologies and innovative solutions, enhancing maritime operations' sustainability (Becker et al., 2013). Academic and research institutions further bolster these efforts by contributing scientific insights that inform evidence-based decision-making (Poo & Yang, 2024, Poo et al., 2021, Yang et al., 2015, 2018). International organisations foster global cooperation, aligning strategies and best practices to address maritime resilience on a broader scale (Akpinar & Özer-Çaylan, 2023). Lastly, communities are key stakeholders, as they actively participate in adaptation initiatives and provide essential local knowledge to address region-specific risks (Camare & Lane, 2015). This multi-stakeholder collaboration is vital to minimising climate impacts across different scales, from local ports to global supply chains, ensuring a cohesive and practical approach to resilience in maritime transport.

5 Conclusion

Maritime transport systems face significant climate adaptation challenges that require a proactive and integrated approach. Implementing infrastructure upgrades, such as improved drainage systems, breakwaters, and elevated port designs, is crucial to counteract the impacts of rising sea levels, extreme weather events, and flooding. Technological innovations, including real-time flood monitoring and thermal-resistant materials, further enhance the resilience of ports and their interconnected systems. Equally important is the role of stakeholder collaboration. Port authorities, local and national governments, private sectors, academic institutions, international organisations, and communities must work together to ensure the success of these adaptation measures. Each stakeholder brings unique strengths—from funding and policy-making to scientific research and local knowledge—making their cooperation indispensable in addressing the multifaceted challenges of climate change. This chapter highlights the urgency of adopting a multi-disciplinary and collaborative approach. Only through cohesive efforts can maritime transport systems build resilience, safeguard global supply chains, and navigate the uncertainties of an evolving climate landscape.

References

Akpinar, H., & Özer-Çaylan, D. (2023). Organisational resilience in maritime business: a systematic literature review. *Management Research Review, 46*, 245–267.

Becker, A. H., Acciaro, M., Asariotis, R., Cabrera, E., Cretegny, L., Crist, P., Esteban, M., Mather, A., Messner, S., & Naruse, S. (2013). A note on climate change adaptation for seaports: a challenge for global ports, a challenge for global society. *Climatic Change, 120*, 683–695.

Becker, A., Ng, A. K., Mcevoy, D., & Mullett, J. (2018). Implications of climate change for shipping: Ports and supply chains. *Wiley Interdisciplinary Reviews: Climate Change, 9*, e508.

Camare, H. M., & Lane, D. E. (2015). Adaptation analysis for environmental change in coastal communities. *Socio-Economic Planning Sciences, 51*, 34–45.

Chen, S. S., Tsang, D. C., He, M., Sun, Y., Lau, L. S., Leung, R. W., Lau, E. S., Hou, D., Liu, A., & Mohanty, S. (2021). Designing sustainable drainage systems in subtropical cities: Challenges and opportunities. *Journal of Cleaner Production, 280*, 124418.

Chhetri, P.K. and Cairns, D.M., (2016). Dendroclimatic response of Abies spectabilis at treeline ecotone of Barun Valley, eastern Nepal Himalaya. *Journal of Forestry Research*, 27(5), pp.1163–1170.

Hughes, J., Cowper-Heays, K., Olesson, E., Bell, R., & Stroombergen, A. (2021). Impacts and implications of climate change on wastewater systems: A New Zealand perspective. *Climate Risk Management, 31*, 100262.

Mastrandrea, M. D., Field, C. B., Stocker, T. F., Edenhofer, O., Ebi, K. L., Frame, D. J., Held, H., Kriegler, E., Mach, K. J., & Matschoss, P. R. (2010). *Guidance note for lead authors of the IPCC fifth assessment report on consistent treatment of uncertainties*. IPCC.

Mcintosh, R. D., & Becker, A. (2019). Expert evaluation of open-data indicators of seaport vulnerability to climate and extreme weather impacts for US North Atlantic ports. *Ocean & Coastal Management, 180*, 104911.

Mikaelsson, M. A., & Dzebo, A. (2023). *Impacts and risks from climate change on trade infrastructure*. Stockholm Environment Institute.

Ngamkhanong, C. (2021). *Resilience of ballasted railway tracks exposed to extreme temperature*. University of Birmingham.

Palin, E. J., Stipanovic Oslakovic, I., Gavin, K., & Quinn, A. (2021). Implications of climate change for railway infrastructure. *Wiley Interdisciplinary Reviews: Climate Change, 12*, e728.

Poo, M. C.-P., & Yang, Z. (2024). Optimising the resilience of shipping networks to climate vulnerability. *Maritime Policy & Management, 51*, 15–34.

Poo, M. C.-P., Yang, Z., Dimitriu, D., Qu, Z., Jin, Z., & Feng, X. (2021). Climate change risk indicators (CCRI) for seaports in the United Kingdom. *Ocean & Coastal Management, 205*, 105580.

Rincon, L. F., Moscoso, Y. M., Hamami, A. E. A., Matos, J. C., & Bastidas-Arteaga, E. (2024). Degradation models and maintenance strategies for reinforced concrete structures in coastal environments under climate change: A review. *Buildings, 14*, 562.

Sanchez, L., & Reames, T. G. (2019). Cooling Detroit: A socio-spatial analysis of equity in green roofs as an urban heat Island mitigation strategy. *Urban Forestry & Urban Greening, 44*, 126331.

Sasidharan, M., Parlikad, A. K., & Schooling, J. (2022). Risk-informed asset management to tackle scouring on bridges across transport networks. *Structure and Infrastructure Engineering, 18*, 1300–1316.

Schulz, A., Zia, A., & Koliba, C. (2017). Adapting bridge infrastructure to climate change: institutionalising resilience in intergovernmental transportation planning processes in the northeastern USA. *Mitigation and Adaptation Strategies for Global Change, 22*, 175–198.

Song, D. (2021). A literature review, container shipping supply chain: Planning problems and research opportunities. *Logistics, 5*, 41.

Yang, Z., Cahoon, S., Chen, S.-L., & Ng, A. K. (2015). *Analysing risks posed by climate change on ports: A fuzzy approach*. Routledge.

Yang, Z., Ng, A. K., Lee, P. T.-W., Wang, T., Qu, Z., Rodrigues, V. S., Pettit, S., Harris, I., Zhang, D., & Lau, Y.-Y. (2018). Risk and cost evaluation of port adaptation measures to climate change impacts. *Transportation Research Part D: Transport and Environment, 61*, 444–458.

Young, A. F., Marengo, J. A., Coelho, J. O. M., Scofield, G. B., De Oliveira Silva, C. C., & Prieto, C. C. (2019). The role of nature-based solutions in disaster risk reduction: the decision maker's perspectives on urban resilience in São Paulo state. *International Journal of Disaster Risk Reduction, 39*, 101219.

Zhou, Q., Panduro, T. E., Thorsen, B. J., & Arnbjerg-Nielsen, K. (2013). Adaption to extreme rainfall with open urban drainage system: An integrated hydrological cost-benefit analysis. *Environmental Management, 51*, 586–601.

E-commerce and Resilient Supply Chain Management During the COVID-19 Pandemic

1 Introduction

The emergence and impact of the COVID-19 pandemic have caused the stagnation of global economic development, but also, provided a major opportunity for international trade to advance its digital transformation for the rise of e-commerce sales. Indeed, social distancing has determined the growing tendency towards online consumption, and e-commerce has become an important way to connect manufacturing with global consumption. In response to this opportunity, there is an acute need to quickly adapt to new supply-demand relationships worldwide, and thus, this book chapter shows the urgent demand for industrial practitioners, digital expertise, policymakers, government bodies, and local communities to understand the current development and future direction of e-commerce businesses, for instance, the benefits of e-commerce, the implementation of e-commerce applications, the use of innovative technologies and knowledge management, research knowledge and skills, education and training, regulatory and governance techniques, market opportunities and winning strategies of e-commerce.

The pandemic has caused the stagnation of global economic development but also provided a major opportunity for international trade to advance its digital transformation. As people turn more to online consumption during the pandemic, cross-border e-commerce has become an important way to connect manufacturing with global consumption via the Greater Bay Area. As the most concentrated area of cross-border e-commerce in China, the Greater Bay Area has brought together world-class financial services, technological innovation and high-end manufacturing centres. The growth of digital trade in the Greater Bay Area has also been rapid. In 2020, the total international trade volume of the Greater Bay Area exceeded HK$15 trillion (CPIT, 2021). It is estimated that in 2035, 50% of the foreign trade volume will be completed through cross-border e-commerce. With the further integration of the economy of the Greater Bay Area, business exchanges in the region

Y.-y. Lau et al., *Maritime Transport and Supply Chain Resilience*,
https://doi.org/10.1007/978-3-032-07566-6_9

have become more frequent and closer, and the demand for financial services such as cross-border payment and trade financing has grown significantly. At the same time, the digital transformation of enterprises in the region has been accelerating during the pandemic, and cross-border e-commerce has become an important channel for expanding foreign trade. As an important international financial centre in the region, Hong Kong is the preferred offshore business centre for mainland Chinese e-commerce companies and can become an important hub to connect companies in the Greater Bay Area with the global market. The role of blockchain technology in promoting international trade is prominent. Paperless logistics based on digital technology can facilitate trade flow and ensure the on-time delivery of goods; more importantly, digital systems and related regulations help to provide safe and reliable payments and financial transactions, thereby increasing market liquidity and reducing financial cost risks. To better support enterprises in the region to expand digital trade, the governments and financial institutions in the three regions of the Greater Bay Area are further increasing their investment in innovative technologies such as big data, artificial intelligence and blockchain, in anticipation of connecting the three regions through financial technology. The financial market will further enhance the facilitation of cross-border trade in the Greater Bay Area.

In general, this chapter is divided into five sections. Section 1 provides research settings and background. Then, we conduct a literature review in Sect. 2. In Sect. 3, the research methodology covers the research context, research design, and the participants. Findings and discussion are given in Sect. 4. Then, we address the summary notes in Sect. 5.

2 Literature Review

Epidemiological modelling provides useful insight for planning and mitigation against possible disasters. However, the inadequacy of academic research on epidemic planning and implementation for rapidly spreading large-scale epidemics remains a major research gap. Dasaklis et al. (2012) proposed further logistics research opportunities: multidisciplinary synergies, end-to-end approaches, more realistic pandemic supply chain planning, inventory replenishment policies, quantification of contingency plans, cross-functional drivers, and coordination issues.

However, we identify two major deficiencies: (1) An effective information-sharing approach is lacking under lockdown policy (e.g. 'How do Hongkonger lockdown cities/borders share information?'). Sharing information enables business supply chains to adopt the appropriate policy, ultimately matching supply and demand (e.g. surgical masks and sanitisers). The supply chain comprises a network of information and epidemic impacts bear out differently for different players along supply chains. Often focusing on local experiences, it is difficult for isolated research to identify issues that can easily transfer between different cultures and countries, not to mention solutions that can catalyse collaboration and information sharing between different cities, countries, and stakeholders (e.g. health officials) in pandemic mitigation planning. Such diversity significantly affects the attitudes of policymakers in pandemic supply chain planning, and hence information management in pandemic control supply chain is vital. (2) The literature lacks an integrated approach for the adoption of epidemic supply chain. Pharmaceutical interventions should be combined with non-pharmaceutical ones (Dasaklis et al., 2012). For example, school closure and travel restrictions can be combined with a pandemic supply chain planning approach. More comprehensive and integrated approaches are needed to embrace supply chain operations from sourcing medical supplies and supplementary commodities to dispensing them to affected populations (e.g. 'how to dispense commodities to a lockdown city'). Also, an investigation of an integrated approach helps to evaluate centralised and decentralised policies of inventory management. Finally, it calls for the involvement of the general public. The reduction or termination of transportation service, spreading rumours, and stocking (e.g. surgical masks, sanitisers) require the general public to cooperate and get involved in pandemic supply chain management. Such negligence means that the solutions from existing research are not adaptive in pandemic supply chain planning.

3 Methodology

3.1 Research Context

We conduct a case study in Hong Kong. Hong Kong has been selected since it has world-leading supply chain systems and is an international trading centre. Also, Hong Kong is strategically situated in the heart of Asia-Pacific and has long been acclaimed for its booming supply chain industry and trading activities. Additionally, establishing cross-border e-commerce businesses aligns with the agreement signed by Chinese President Xi Jinping, the National Development and Reform Commission, and the Guangdong, Hong Kong, and Macao governments, that is, the Framework Agreement on Deepening Guangdong-Hong Kong-Macao Cooperation in the Development of the Greater Bay Area. Our book chapter can increase the awareness and support from the industries to share ways to facilitate regional economic development and encourage greater economic integration within the Greater Bay Area via cross-border e-commerce business development.

In general, Hong Kong has undergone five waves of the COVID-19 pandemic from January 2020 to July 2022. In the past 2.5 years, Hong Kong has accumulated several infected people from the number of confirmed cases of COVID-19. In response, the HKSAR government has implemented four social restriction measures to control the transmission of the virus during the outbreak of the COVID-19 pandemic. The summary of the COVID-19 pandemic that occurred in Hong Kong is shown in Fig. 1.

How Hongkonger lockdown cities facilitate valid information sharing remains unclear. This implies that Hong Kong, and even the whole world, urgently needs a

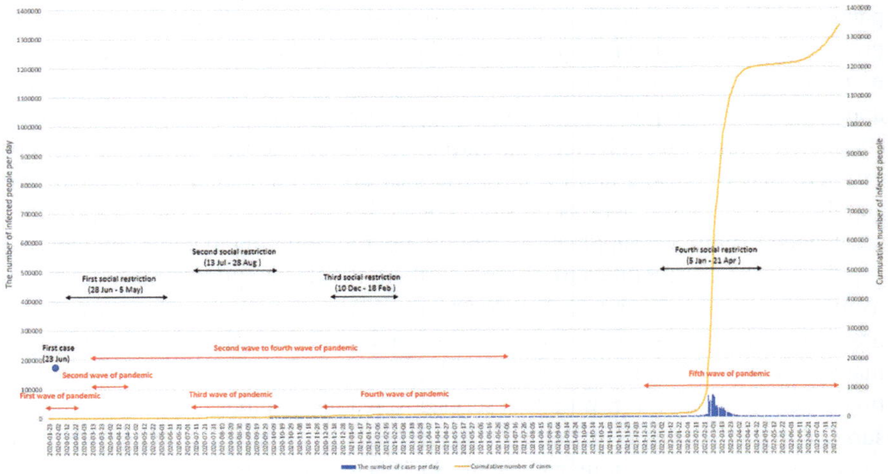

Fig.1 Summary of COVID-19 pandemic cases in Hong Kong

new supply chain planning system during a pandemic, and our outputs provide valuable insight into how such a new system should be designed, established, and implemented.

3.2 *Research Design*

The survey questions and key information in the discussion section were based on previous research studies and semi-formal in-depth discussions with various parties or materials containing researchers, industrial practitioners, and logistics associations. To validate and align with the study objectives, the questionnaire has mainly removed vague wordings and cleared double-barrel items. Most areas of the items can be measured and validated in the amended final survey questionnaire. It is vital to ensure the validity of their content, which is a crucial measure of survey instruments' correctness. Examining content validity generally contains a systematic review of the survey's content to ensure that we eliminate everything it should not and incorporate anything it should (Lau et al., 2018).

The sample of our survey is the public which involves opinions and reactions towards shortage of supplies or commodities due to the lockdown. Survey results can accurately identify and measure people's attitudes, reactions, and fears. The survey consists of two main sections (41 close-end questions). The first section documents general respondent characteristics in terms of age, geographic location, and job title, to name but a few. The second section documents the attitude and perception of respondents on the pandemic supply chain planning, observed changes, and response policies employed. Specific questions aim to (1) identify pandemic-related supply chain risks perceived, (2) evaluate the observed impact of a pandemic on commodities and over-the-counter medical supplies, (3) document pandemic policy adopted by Hong Kong organisations, and (4) record individuals' response to the pandemic supply chain. Understanding that pandemic supply chain planning is an embryonic topic with few preceding policies and solutions gives us an ideal opportunity to validate the explanatory power of the policy implementation theory and identify ways that improve policy implementation theory under new contexts.

3.3 Participants

An online survey has been conducted in Hong Kong in March 2022 via a convenience sampling method. Finally, we have received 410 valid responses from the target respondents. The survey respondents' answers are only used for academic purposes and are highly confidential in this research study. Thus, the responses gathered will only be reviewed in aggregate. Table 1 exhibits the survey respondents' profiles. Over half of the participants were female (52.44%). 84.63% of survey respondents were aged between 18 and 35 years old. Over half of the participants indicated living in the New Territories (52.68%). Regarding educational level, 54.15% of survey respondents obtained a bachelor's degree or above.

Table .1 Profile of survey respondents

Demographics	Number of respondents	Percentage
Gender		
Male	195	47.56
Female	215	52.44
Age		
18–25 years	303	73.90
26–35 years	44	10.73
36–50 years	41	10.00
51–65 years	20	4.88
Over 65 years	2	0.49
Living location		
Hong Kong Island	45	10.98
Kowloon	142	34.63
New territories	216	52.68
Outlying Island	7	1.71
Educational level		
Sub-degree or below	188	45.85
Undergraduate	163	39.76
Postgraduate	59	14.39

4 Findings and Discussions

How often did you shop in a physical store before COVID-19, such as purchasing life's necessities and other products in the public market or supermarket? (Table 2)

What kind of products do you spend the most before and during COVID-19? (Table 3)

How often did you shop online during COVID-19? (Table 4)

Where do you approach the commodity information during COVID-19? (Table 5)

What is the concern of different age groups during COVID-19? (Table 6)

Descriptive analysis and ANOVA were conducted on survey data gathered from 410 Hong Kong residents. Before the COVID-19 pandemic, 73.5% of respondents indicated that they would purchase life's necessities and other products in a physical store. However, only 20.5% of survey respondents expressed that they did shop online during COVID-19. Regarding the survey participants who spent the most on products before and during COVID-19, the top three shopping lists were food, dining and lifestyles, followed by housewares and personal care and health products before COVID-19, whereas the top three shopping lists were food, dining and lifestyles, followed by disease prevention product and housewares during COVID-19. In addition, the survey participants identified physical stores as the most effective way to obtain commodity information during COVID-19. Furthermore, the elderly performed weaknesses in identifying fake information about the commodity information and engaging in online shopping due to their attitudes, knowledge and experience. Besides, the general public responded to the different logistical needs in pandemic outbursts at different stages via content analysis.

Table 2 The frequency of survey participants shop in a physical store before COVID-19

Item	Number	Percentage
Never	8	1.8
Once a month	48	11.0
Twice a month	60	13.7
Once a week	115	26.3
Twice a week	88	20.1
Three times or more a week	119	27.2

Table 3 The comparison of products that survey respondents spend before and during COVID-19

Item	Before—Number	Before—Percentage	After—Number	After—Percentage
Personal care and health product	53	12.1	58	13.3
Skin care and make-up	23	5.3	15	3.4
Electrical appliances	13	3.0	13	3.0
Housewares	75	17.2	66	15.1
Food, dining and lifestyles	231	52.9	163	37.3
Sports and travels	8	1.8	1	0.2
Toys and books	7	1.6	11	2.5
Fashion and accessories	19	4.3	20	4.6
Insurance and finance	1	0.2	2	0.5
Disease prevention product	7	1.6	88	20.1

Table 4 The frequency of survey participants shop online during COVID-19

Item	Number	Percentage
Never	115	26.5
Once a month	143	32.6
Twice a month	89	20.3
Once a week	34	7.8
Twice a week	35	8.0
Three times or more a week	21	4.8

Table 5 The possible ways for survey participants to collect commodity information during COVID-19

Item	Number	Percentage
Physical store (e.g. supermarket)	149	34.7
Online shop (e.g. HKTV mall)	126	29.4
Website (e.g. Price.com.hk)	83	19.3
Government (e.g. online price watch by consumer council)	9	2.1
Information from friends	62	14.5

Table 6 Issues concerned by different age groups during COVID-19

Item	Mean by age group			Grand mean (SD)	ANOVA-test	Comparison
	18–25 ($n = 145$) (1)	26–50 ($n = 180$) (2)	>50 ($n = 113$) (3)			
There was some fake information about the commodity information, you would fact-check product information to identify the information's accuracy.	3.70	3.66	3.31	3.58 (0.84)	8.17**	1 > 3, 2 > 3
You think you enjoy high information transparency during COVID-19 that you could approach all information and detail about the product when you search about the product and brand in internet.	3.28	3.28	3.18	3.26 (0.78)	0.76	
The product transportation duration is longer during COVID-19.	3.68	3.63	3.40	3.59 (0.98)	2.99	
Do you think tracking product position during COVID-19 enables to increase accuracy the product delivery time?	3.42	3.46	3.19	3.38 (0.91)	3.09*	2 > 3
After the outbreak, you tend to purchase a quantity of life's necessities in a single shopping online.	3.03	3.24	2.82	3.07 (1.02)	6.16*	2 > 3
After the outbreak, you tend to purchase a quantity of life's necessities in a single shopping trip through physical shopping.	3.44	3.41	3.16	3.35 (1.00)	2.91	
Online shopping could lower the infection risk from COVID-19	3.78	3.62	3.41	3.62 (0.84)	6.28**	1 > 3

5 Conclusion

Cross-border e-commerce has registered an upward trend, and in the pandemic context, it has soared. This book chapter wants to shed light on the industry and business implications related to cross-border e-commerce opportunities and challenges. Although sales increase, businesses may face unpredictable challenges like product availability, logistics and transportation disruptions, and consumer protection to avoid fraudulent and dishonest practices. To respond to these opportunities and challenges, it implies that industries and businesses need to be strategic and find ways to deal with these evolving issues. For instance, e-commerce firms may have to adapt their business models, introduce and expand their products and services, and adopt new technologies.

This groundbreaking study causes a paradigm shift in the research approach in the study of pandemic supply chain planning. While decision-makers and stakeholders nowadays manage to propose pandemic supply chain solutions, they often find it difficult to move to the next stage and effectively implement such policies and plans due to information sharing and lack of an integrated approach (Dasaklis et al., 2012; Adivar & Selen, 2013). This study generates vital information on how researchers, policymakers, and the general public can easily transfer knowledge to cities or countries and demonstrate how the supply chain response and decision-making process to sudden, large-scale pandemic outbreaks, such as COVID-19, can be effectively improved. In turn, this encourages collaboration between countries and cities to improve the quality of decision-making. As the general public, such as the stakeholders, participates in exploring an emerging research area, this study offers scholars, practitioners, policymakers and other stakeholders a rare opportunity to work together and thoroughly understand how to collaborate and develop policies effectively. This book chapter's findings are highly transferred to other sectors. As supported by real-world evidence, large-scale infrastructures (e.g. airports, metro, etc.) in Hong Kong and around the world often face similar challenges in dealing with pandemic outbreaks. The decision-makers of these facilities find this study extremely helpful when developing similar mitigation policies and measures. Finally, it helps Hong Kong and the world to get out of its current impasse by offering valuable insight into the design, establishment, and implementation of its new governance system.

References

Adivar, B., & Selin Selen, E. (2013). Review of research studies on population specific epidemic disasters. *Disaster Prevention and Management: An International Journal, 22*(3), 243–264.

Council for the Promotion of International Trade (CPIT) of the Greater Bay Area (GBA). (2021). Retrieved January 11, 2022, from http://www.dg.gov.cn/dgsmch/gkmlpt/content/3/3696/post_3696625.html#1540

Dasaklis, T., Pappis, C., & Rachaniotis, N. (2012). Epidemics control and logistics operations: A review. *International Journal of Production Economics, 139*(2), 393–410.

Lau, Y. Y., Ng, A. K. Y., Tam, K. C., & Chan, E. K. K. (2018). An investigation on the professionalization of education in maritime logistics and supply chains. *Maritime Business Review, 3*(4), 394–413.

A Case Study on the Impacts of a Pandemic on the Resilience of Supply Chains and Regional Economy

1 Introduction

In December 2019, a surge of serious acute normal respiratory diseases happened in Wuhan, China. This was a doubtful viral disease prior to December 2019. The most general symptoms of this virus are fever, dry cough and tiredness. An epidemic of SARS-CoV-2 was identified as a zoonotic spreading arising from a Huanan seafood wholesale market in Wuhan, China. Indeed, human-to-human transmission significantly boosted the spread of the virus. The disease created by the virus is typically recognised as COVID-19 (Lau et al., 2022). COVID-19 has quickly transmitted to different parts of the globe in a moment due to antiviral treatments being unavailable or no particular vaccines yet except for trials (e.g. hydroxychloroquine) (Lau et al., 2020). As such, the World Health Organization (WHO) announced COVID-19 as a worldwide pandemic, which is indicated as 'the occurrence of disease cases in excess of normal expectancy'. As Yu et al. (2020) addressed that it induces a global crisis and emerges as a pandemic. By 2022, COVID-19 has caused the deaths of 6,459,684 people around the world and 43,505 in Canada (WHO, 2022).

The rate of unemployment around the world has skyrocketed and unemployment claim has never been as high as now. The impact of COVID-19 on the for-profit sector in Canada has been significant as evidenced by high unemployment rates and the increasing number of businesses filing cases for bankruptcy (e.g. HERTZ). In this case, the province of Manitoba (hereinafter called 'Manitoba') has the highest rate of COVID-19 confirmed cases across Canada. The location of Manitoba is provided in Fig. 1. The rate of confirmed cases in Manitoba is 1,538 per 100,000 Manitobans (CTV News Winnipeg, 2022). The main reasons are (1) Manitoba was relatively slow to perform extra measures to deploy additional hospital resources; (2) Manitoba includes various other people of colour or Indigenous people who either have poor access to health-care services or poor health; (3) northerners and rural Manitoba have restricted access to health care because the health care services

Y.-y. Lau et al., *Maritime Transport and Supply Chain Resilience*,
https://doi.org/10.1007/978-3-032-07566-6_10

are highly concentrated on the provincial south and Winnipeg; and (4) Manitoba has recorded a low vaccination rate (CBC, 2022).

According to a survey during the pandemic, 12% of Manitoba's small businesses are considering filing for bankruptcy (Malak Abas, 2020). Moreover, the continuous closure of the US-Canada border to non-essential travel puts a significant percentage of the GDP of the Canadian provinces at risk. For Manitoba, 39% of its GDP is tied to the USA (Jones, 2020). Even though the major exports from Manitoba to the USA are agriculture and machinery (Manitoba Bureau of Statistics, 2019), the border closure has indirectly affected the province's other key sectors as well, like tourism. Further, the airline industry has virtually come to a standstill, laying off an unprecedented number of staff. For example, Air Canada and WestJet are laying off 20,000 and 3333 employees, respectively (George-Cosh, 2020; CBC/Radio-Canada, 2022). Boeing is laying off as many as 400 people in its Winnipeg plant (Caitlyn, 2020). This is a huge blow to Manitoba's vibrant aerospace industry. Other airplane parts manufacturing companies are likely to follow suit. The energy sector is also laying off employees, such as Ovintiv is laying off about 25% of its staff (Healing, 2020). Although a recent survey conducted by the Manitoba Chamber of Commerce reveals that more than two-thirds of businesses did not close during the pandemic, there are some that were hit rather hard, especially tourism-related businesses (Manitoba Chamber of Commerce, 2020).

In this chapter, we discuss a historical case identifying the immediate impacts of COVID-19 on the for-profit sector in Manitoba specifically, but also in Canada, generally, between 2020 and 2022. Also, we provide constructive analysis of how governments should proceed with the recovery, including the enhancement of supply chain resilience.

2 Canada and Manitoba COVID-19 Pandemic-Related Laws

As part of the efforts by the Manitoban Government to help its citizens to cope with the financial aspects of the pandemic, Finance Minister Scott Fielding announced deferred tax filing deadlines, postponed eviction hearings, and rent freeze programs to help for-profit businesses. The filing deadline has been postponed to September 30, 2020 (Canada Revenue Agency, 2020). These measures are geared towards reducing the burden on business.

Furthermore, the Manitoba Public Insurance (MPI) has provided relief for clients, which includes individuals and for-profit businesses, within this period. It provided refunds worth approximately 5% of the total amount paid within the 2018/2019 and the 2019/2020 years. Another important step taken by the Manitoban government is the Manitoba Gap Protection Program (MGPP), where up to CAD120 million is provided to support small- and medium-sized businesses that do not meet the criteria for most of the federal government programs (Manitoba Government, 2020a).

To support students (and businesses), the Manitoban Government has made available approximately CAD120 million in the form of subsidies to employers if they hired students between May 1 and September 4, 2020. For online shopping, the government (provincial and federal) has provided CAD160,000 to farmers and local producers to facilitate business transactions online. Finally, the trucking industry received a boost from the Government granting them an extension of expiry dates on certain permits, certificates and/or registrations. The industry had until July 3, 2020, to fulfil this obligation (Manitoba Government, 2020b). Most deadlines imposed have since been extended (Lara et al., 2020) and even though as of the time of conducting this study, there has not been an update on the extension for the trucking industry, generally tax and fee deferrals have been extended (Manitoba Government, 2020c).

Furthermore, Bill C-14, the COVID-19 Emergency Response Act, No. 2 (CERB) has been launched in Manitoba. It seeks to provide relief to employers in terms of complementing wages up to approximately 75%. The Canadian federal government has set aside about 40 billion CAD to provide loans to companies. Nevertheless, key challenges arise from fraud applications, and in response, the proposal to remedy the situation by parliament. Recipients are to pay up to double what they received, and CAD5,000 fine, or six-month imprisonment and a CAD5,000 fine (Parliament of Canada, 2020).

3 Panic Buying

In March 2020, a single roll of toilet paper was going for as much as approximately CAD10 in Winnipeg, while a 12-roll package was as expensive as CAD1,000 (Marek and Elisha, 2020). Hoarding of essential items was investigated in Winnipeg. By April 2021, the price stood at about CAD20 for a 30-pack. Figure 1 shows the

Fig. 1 An example of price hoarding of essential medical goods (currency in Canadian Dollars; photo on the left taken by the authors in a grocery store in Winnipeg, Manitoba, Canada, April 2020). Source: Authors

Table 1 The most commonly used information source by Manitobans

Items	Percentage of responses
Government's public information	17.95%
Mass media including news, radio, magazine, television, and social media	69.23%
Reference group: Family members, friends, colleagues, neighbours	5.13%
Personal experience or had a similar experience	2.56%
Skip	2.56%
Other	2.56%

Source: Authors

cost of disposable face masks during this period. While they cost CAD54 in the early phases of COVID-19, by April 2021, they went for USD17.99 (CAD24.08) on Amazon.com. Amazon.ca did not have the Modenna brand on the site as of the time of writing. Such price gouging behaviours hurt everyone, especially the vulnerable and low-income earners. However, it is noted that not every business in the for-profit sector experienced price gouging.

Regarding panic buying, the authors collected survey data from 392 Manitobans between March and May 2021, a time when the province was still in the midst of experiencing the impacts and challenges. The data collection channels included both physical and online means. In this case, 'Manitobans' were understood as individuals who indicated that Manitoba was their place of residence in the past 12 months, including students. The most commonly used information source by Manitobans is summarised in Table 1. According to the collected data, Manitobans mainly relied on mass media like news, radio, magazine, television, and social media, followed by the Canadian and/or Manitoban governments' public information and reference group including family members, friends, colleagues and neighbours. In response, most Manitobans reflected that they trust the information source (3.51) (i.e. 1 = strongly disagree; 5 = strongly agree) due to less than 3% of respondents having a similar experience or personal experience.

For hoarding items, the number of hoarding items and the reasons for not hoarding is summarised in Tables 2 and 3, respectively. In terms of hoarding items, groceries, hand sanitiser, medical masks, and toilet paper were identified as the most useful epidemic prevention supplies from the Manitobans' perspectives. However, the over-the-counter drug seems an ineffective way to consider specific antiviral treatments for COVID-19. Interestingly, the participants pointed out that the location of convenience stores or supermarkets is one of the main reasons for not hoarding. Indeed, the participants inclined a passive attitude towards the seriousness of the COVID-19 pandemic. Indeed, almost 55% of participants perceived that it was not necessary for hoarding.

Manitobans' shopping frequency and shopping approach for groceries during the pandemic, the results are presented in Tables 4 and 5, respectively. Largely in-store shopping remained the mainstream shopping approach for groceries during the pandemic. The main reasons were residents in northern Canada and rural Manitoba had

Table 2 The number of hoarding items

Hoarding items	Percentage of responses
Medical masks	46.88%
Groceries	50.00%
Over-the-counter drug	6.25%
Toilet paper	31.25%
Hand sanitiser	50.00%
Other	25.00%

Source: Authors

Table 3 Reasons for not hoarding

Items	Percentage of responses
Convenient store/supermarket nearby	27.27%
No storage space at home	6.06%
Online shopping	9.09%
Do not consider it necessary	54.55%
Other	3.03%

Source: Authors

Table 4 Shopping approach for groceries during the pandemic

Items	Percentage of responses
In-store shopping	65.00%
Online shopping	32.50%

Source: Authors

Table 5 Shopping frequency of groceries during the pandemic

Items	Percentage of responses
Less than usual	51.28%
Usual	38.46%
More than usual	10.26%

Source: Authors

rather limited internet access; indigenous and elderlies were generally not well-educated or possessed limited knowledge of buying groceries via a (from their perspective) complicated online system and payment method. Due to the poor-quality transport infrastructure, it led to over half of Manitobans performing less than usual in shopping frequency of groceries during the pandemic.

Furthermore, many Manitobans expressed that they were actually not very anxious about the impacts of the COVID-19 pandemic (3.36) (i.e. 1 = not at all anxious; 5 = extremely anxious) and did not feel extremely threatened by COVID-19 (3.44) (i.e. 1 = not at all threaten; 5 = extremely threaten). As such, it could perhaps provide a convincing explanation on why panic buying was not that serious in Manitoba in response to COVID-19.

4 Border Closure with the USA and Provinces

Tourism was a key industry that was hit hard by the closure of services in Canada, and Manitoba was no exception. According to the Manitoba Lodges and Outfitters Association, just before the pandemic, visitors from the USA represented approximately 90–95% of their customers. Another area that has been seriously affected was Churchill, where beluga whale watching (a major source of revenue for the town and its surrounding regions during the early summer) has been seriously impacted (Josh, 2020). Further, border towns like Emerson-Franklin deeply felt the pain of the closure. Some border towns were so intertwined with their USA counterparts that even emergency response came mainly from the other side of the border (Sam, 2020). All such had clearly disrupted the smooth flows of transport and supply chains.

5 Lessons Learnt

While Manitoba had the bulk of its GDP dependent on the USA, the silver lining was that it also offered the province an opportunity to diversify in terms of the destination of exports. Furthermore, to overcome the reduction in tourism, the Manitoban government might have to liaise with the Canadian government to introduce a strict quota monitoring regime for visitors, to boost the tourism sector, as opposed to no visitors at all.

Moreover, the disruption highlighted the importance for Manitoba to devise a way to attract highly skilled immigrants and top talents to the province, as the pandemic necessitated the need for the province to undergo some fundamental changes in its economic activities and structures. Recessions are great times to start companies and it is extremely difficult to get the best brains in a particular city during normal periods, especially for an inland, regional center like Winnipeg. However, anytime that happens, innovation, and entrepreneurship takes place; therefore, many new firms have been established. Looking to the future, Tthere is a strong need to increase research funding both at universities and companies. As such, it accelerated a ‘Made in Manitoba’ initiative to make sure that most of the things the province needs are produced locally. There should be highly favourable tax breaks to Manitoba-based companies and a more favourable tax policy that would attract some of the greatest technology companies to come to Winnipeg or even other cities and towns in Manitoba. Amazon, Google, and other giants often lobby for this and so the Manitoban government needs to get some of these companies to the province. This way, it would create more opportunities for young professionals as well as other accomplished workers, thus encouraging them to come and stay in Manitoba.

Further, during the pandemic the province opened more to new investments, including the supply chain sector. For example, the telecom industry is one that has substantial potential. Finally, there is a need to bring major manufacturing back to

Manitoba, especially for essentials like health care and basic household items. Indeed, with the latest announcements of the Manitoban government on the development of new partnership for CentrePort Canada's rail park (CentrePort Canada, 2022), it creates a new opportunity for Manitoba to improve its logistical, including maritime, connections with other parts of the world, as well as attracting new investments to the province. The long-term impacts on Manitoba is subject to further research and investigation.

6 Conclusion

The COVID-19 pandemic, including the damage and impacts, have made individuals, organisations and governments take a second look at how they should run businesses and economies. That said, it is an ideal time to be innovative and develop new business models as demonstrated in Hong Kong and Mainland China during the SARS pandemic in the early 2000s. Manitoba was not spared by the COVID-19 pandemic. In fact, the economy contracted in the first quarter of 2020 and a lot of workers were laid off both by the Manitoban government and for-profit organisations with the goal of minimising costs. The Manitoban government enacted laws to protect for-profit organisations, address unemployment and manage rents for individuals and businesses. The laws enacted by the Manitoban Government have closed some of the loopholes of the federal laws which, among others, have closed the Canada-USA border for a prolonged period of time. The consequence of this closure is reduced growth in the tourism sector and among like businesses. Looking at the future, the for-profit sector in Manitoba would take a while to recover from the COVID-19 pandemic. Nevertheless, with the right approach, the recovery process could be faster, smoother and less painful. We believe that this study offers some useful insight into how this can be achieved in the foreseeable future. It is an ideal time to be innovative and create new companies, as demonstrated in Hong Kong and China during the SARS pandemic in the early 2000s. Countries that were able to deal with the pandemic effectively also face economic consequences, vice versa. Where is the priority? Economy or health? How do we find the middle ground when shipping, logistics and global supply chain are intertwined, hence enhancing the sector's resilience? All these require further investigations and research, and we believe that the case of Manitoba offers a decent platform for this.

References

Caitlyn, G. (2020). 400 people to be laid off at Boeing in Winnipeg in coming weeks. *CBC News*. Retrived May 24, 2020, from https://www.cbc.ca/news/canada/manitoba/boeing-layoffs-winnipeg-covid-19-1.5582503

Canada Revenue Agency. (2020). *Income tax filing and payment deadlines*. Retrived November 6, 2020, from https://www.canada.ca/en/revenue-agency/campaigns/covid-19-update/covid-19-filing-payment-dates.html

CBC. (2022). *Manitoba has Canada's highest COVID hospitalizatoin rate*. Retrived August 28, 2022, from https://www.cbc.ca/news/canada/manitoba/manitoba-covid-hospitalizations-analysis-1.6321890

CBC/Radio-Canada. (2022). *Air Canada to lay off 20,000 workers as pandemic collapses travel industry*. Retrived August 1, 2022, from https://www.cbc.ca/news/business/air-canada-layoffs-1.5572596

CentrePort Canada. (2022). *Media release: Government of Manitoba announces development partner for the CentrePort Canada rail park*. Retrived September 4, 2022, from https://centreportcanada.ca/media-release-government-of-manitoba-announces-development-partner-for-the-centreport-canada-rail-park/

CTV News Winnipeg. (2022). *Manitoba outpacing rest of Canada in rate of active COVID-19 cases*. Retreived August 28, 2022, from https://winnipeg.ctvnews.ca/manitoba-outpacing-rest-of-canada-in-rate-of-active-covid-19-cases-1.5733938

George-Cosh, D. (2020). *WestJet lays off 3,333 in latest round of aviation job cuts*. Retrived July 26, 2021, from https://www.bnnbloomberg.ca/westjet-lays-off-3-333-in-latest-round-of-aviation-job-cuts-1.1455760

Healing, D. (2020). *Ovintiv confirms 25% of staff laid off amid fears of lower oil demand*. Retrived June 19, 2020, from https://www.bnnbloomberg.ca/ovintiv-confirms-25-of-staff-laid-off-amid-fears-of-lower-oil-demand-1.1452991

Jones, J. (2020). Ovintiv, Formerly Encana, is cutting its work force by 25 percent. *The Globe and Mail*. Retrived June 18, 2020, from https://www.theglobeandmail.com/business/article-ovintiv-formerly-encana-is-cutting-its-work-force-by-25-per-cent/

Josh, C. (2020). *Manitoba's tourism industry bracing for impact of continued Canada-U.S. border closure*. Retrived May 19, 2020, from https://winnipeg.ctvnews.ca/manitoba-s-tourism-industry-bracing-for-impact-of-continued-canada-u-s-border-closure-1.4945968

Lara, N., Trevor, L., Eric, B., Feder, S., & Kara, S. (2020). *COVID-19: Economic relief measures announced to date*. Retrived July 31, 2020, from https://www.mccarthy.ca/en/insights/articles/covid-19-economic-relief-measures-announced-date#MT

Lau, Y. Y., Zhang, Z., Ng, A. K. Y., & Panahi, R. (2020). Implications of a pandemic outbreak risk: A discussion on China's emergency logistics in the era of coronavirus disease 2019 (COVID-19). *Journal of International Logistics and Trade, 18*(3), 127–135.

Lau, Y. Y., Dulebenets, M. A., Yip, H. T., & Tang, Y. M. (2022). Healthcare supply chain management under COVID-19 settings: The existing practices in Hong Kong and the United States. *Healthcare, 10*, 1549.

Malak Abas. (2020). *Manitoba businesses 'reaching a critical point': expert New survey shows 12 percent of small operations considering bankruptcy, winding down*. Retrived August 27, 2021, from https://www.winnipegfreepress.com/business/manitoba-businesses-reaching-a-critical-point-expert-571306152.html

Manitoba Bureau of Statistics. (2019). *Manitoba's strategic advantages: Manitoba trade exports/imports*. Retrived September 28, 2019, from https://www.gov.mb.ca/jec/mbadvantage/business_import_export.html

Manitoba Chamber of Commerce. (2020). *COVID-19 Manitoba business action report*. Retrived June 8, 2020, from https://myemail.constantcontact.com/Your-in-depth-analysis%2D%2DBusiness%2D%2D-COVID-19.html?soid=1117862940050&aid=Pz83hDPh1g

Manitoba Government. (2020a). *Manitoba Gap Protection Program (MGPP)*. Retrived October 3, 2020, from https://www.gov.mb.ca/covid19/protection/mgpp.html

Manitoba Government. (2020b). *Media background on COVID-19 in Manitoba*, available at: https://manitoba.ca/bg/2020/04/covid19.html, accessed on 6 October 2020.

Manitoba Government. (2020c). *Tax and fee deferrals*. Retrived July 28, 2020, from http://www.manitoba.ca/covid19/protection/tax-fee-deferrals.html

Marek, T., & Elisha, D. (2020). *Coronavirus: Panic-buying in Winnipeg brings out re-sellers and claims of price Gouging*. Retrived March 14, 2020, from https://globalnews.ca/news/6673262/coronavirus-panic-buying-in-winnipeg-brings-out-re-sellers-and-claims-of-price-gouging/

Parliament of Canada. (2020). Retrived November 5, 2020, from https://lop.parl.ca/sites/PublicWebsite/default/en_CA/ResearchPublications/LegislativeSummaries/431C14E

Sam, T. (2020). *Manitoba Border Community faces coronavirus-related economic struggles, rising floodwaters*. Retrived April 16, 2020, from https://globalnews.ca/news/6828004/manitoba-border-community-coronavirus-flooding/

World Health Organization (WHO). (2022). Retreived August 27, 2022, from https://covid19.who.int/region/amro/country/ca

Yu, H., Sun, X., Solvang, W. D., & Zhao, X. (2020). Reverse logistics network design for effective management of medical waste in epidemic outbreaks: Insights from the coronavirus disease 2019 (COVID-19) outbreak in Wuhan (China). *International Journal of Environmental Research and Public Health, 17*, 1770–1795.

Best Practices and Guidelines for Resilient Maritime Transport and Supply Chains

1 Introduction

In an era marked by climate change, geopolitical uncertainties, and the need for post-pandemic recovery, resilient maritime transport and supply chains are critical for sustaining global trade flows. This chapter synthesises best practices, practical insights and theoretical advancements in designing and implementing resilience-building measures in the maritime domain. Drawing from real-world examples, the chapter outlines strategies that have successfully enhanced operational robustness, reduced vulnerabilities and maintained continuity in trade operations. These insights are consolidated into an integral resilience model that serves as a framework for guiding the development of sustainable, adaptive and collaborative maritime systems.

2 Key Success Stories

Singapore has set a global benchmark in port efficiency and environmental sustainability through its strategic adoption of automation, digitalisation and green energy initiatives. The port has significantly optimised throughput while minimising its carbon footprint by leveraging cutting-edge technologies such as AI-powered cargo management systems and predictive analytics for vessel scheduling (Datta & Satghare, 2024). These measures underscore Singapore's commitment to sustainable growth and operational excellence in the maritime sector.

The Port of Rotterdam stands out as a leader in climate adaptation, integrating innovative measures to mitigate the risks posed by extreme weather events. Elevated infrastructure, storm surge barriers, and energy-efficient equipment form the cornerstone of its resilience strategy (Taneja et al., 2021). By fostering collaboration

Y.-y. Lau et al., *Maritime Transport and Supply Chain Resilience*,
https://doi.org/10.1007/978-3-032-07566-6_11

with local communities and policymakers, Rotterdam has established a robust framework that safeguards its operations and strengthens its position as a vital hub in global trade.

Japan's response to the 2011 tsunami highlights the importance of rapid recovery and resilience-building in maritime supply chains (Mulopulos, 2023). By incorporating redundancy into supply chain systems and establishing strong public-private partnerships, Japanese ports demonstrated an ability to expedite recovery timelines effectively. These efforts have bolstered preparedness for future disruptions, ensuring the continued stability of Japan's critical maritime infrastructure.

3 Integral Resilience Model for Maritime Transport and Supply Chains

The integral theory by Wilber (2005) is being used as a comprehensive framework for building robust maritime transport and supply chain systems. It incorporates four core dimensions by integrating the details from previous chapters, and the entire integral resilience model is illustrated in Fig. 1.

The Interior Individual dimension focuses on the internal, cognitive and psychological aspects that influence decision-making at the individual level. For shipping executives, this involves managing uncertainty and stress while navigating complex disruptions, such as those caused by climate change or global pandemics. Leadership resilience training and the ability to make rapid, well-informed decisions are key factors of effectiveness in this dimension. Additionally, academia plays a crucial role by contributing research to build resilience frameworks, advancing knowledge

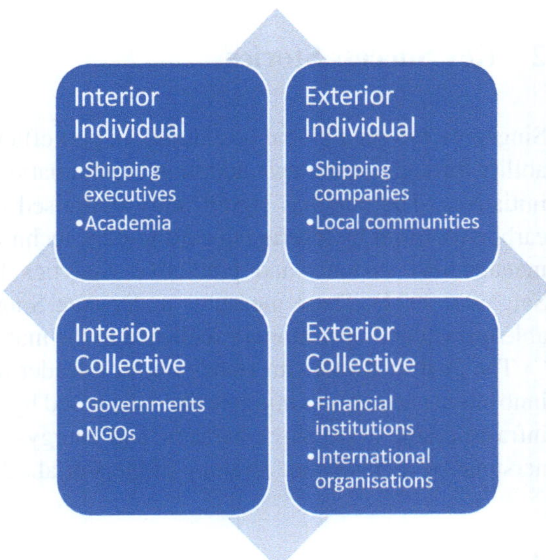

Fig. 1 Integral resilience model for maritime transport and supply chains

and proposing innovative solutions. Metrics such as the number of published papers and their impact in terms of citations highlight the contribution of academic efforts to this dimension.

The Exterior Individual dimension emphasises individuals' behavioural changes and technological adoption to enhance maritime and supply chain resilience. This area includes implementing cutting-edge technologies such as automation, AI and IoT solutions for shipping companies to streamline operations and minimise human error. Automation levels and the reduction of errors serve as indicators of progress. Meanwhile, local communities play a vital role in adopting sustainable practices to support resilience. Behavioural shifts contribute to a broader culture of sustainability. For example, waste reduction efforts are impactful and measured through engagement levels and tangible environmental impacts.

The Interior Collective dimension explores the cultural and organisational resilience within collective entities, such as governments and non-governmental organisations (NGOs). Governments must drive cultural shifts and organisational alignment to prioritise climate-resilient policies, with success measured by regional policy alignment indices and implementation rates. On the other hand, NGOs focus on fostering collective social impact by advocating for vulnerable groups and implementing programs addressing systemic issues. Successful advocacy campaigns and the number of implemented programs provide key metrics to gauge the effectiveness of NGO contributions to collective resilience.

The Exterior Collective dimension highlights the external systemic efforts needed to bolster resilience involving infrastructure, financial systems and international collaboration. Financial institutions are crucial in funding models and providing incentives to promote green shipping and resilient infrastructure. Investments in these areas and completing funded projects are benchmarks for success. International organisations focus on fostering global collaboration through treaties and frameworks that align nations towards common resilience goals. The number of international agreements and adherence rates to these frameworks are indicators of effective coordination and alignment across borders, reinforcing systemic resilience in the maritime and supply chain sectors.

4 Case Study

The Port of Rotterdam exemplifies the integration of these dimensions to enhance maritime resilience. For the Interior Individual dimension, leadership training programs were conducted to equip executives with the tools to navigate disruptions caused by climate change and geopolitical instability (Liu et al., 2023). The port's executives collaborated with academic institutions, such as Erasmus University Rotterdam, to design frameworks for adaptive decision-making, reflected in high-impact publications cited globally.

Rotterdam deployed automation technologies in the Exterior Individual dimension, including predictive analytics for cargo flow management and AI-driven scheduling systems, reducing human error and operational inefficiencies. The port's automation levels increased by 25% between 2018 and 2022, demonstrating measurable progress (De Alwis & Nam, 2024).

Culturally, Rotterdam embodies the Interior Collective dimension by fostering a collaborative ecosystem with policymakers and local communities—government policies like the National Climate Agreement aligned regional goals towards reducing greenhouse gas emissions. (Kin et al., 2024). Simultaneously, NGOs partnered with the port to run programs addressing social vulnerabilities, improving workforce inclusivity and community engagement.

Finally, in the Exterior Collective dimension, the Port of Rotterdam has secured significant investments from financial institutions to fund climate-resilient infrastructure, such as flood-proof docks and renewable energy facilities. It also played a leading role in negotiating international agreements under the International Maritime Organization (IMO), with high adherence rates among partner nations. These efforts underscore Rotterdam's commitment to systemic resilience in global maritime trade (Kok et al., 2024).

This case study illustrates how the four resilience dimensions can be synergistically applied to achieve sustainable and adaptive maritime systems.

5 Conclusion

This chapter consolidates the core themes explored throughout the book, emphasising the criticality of resilience as a cornerstone for sustainable and efficient maritime systems. From the disruptions brought about by the COVID-19 pandemic to the challenges posed by climate change and geopolitical uncertainties, the book emphasises the need to enhance resilience in maritime transport and supply chains. Chapter 1 details the vulnerabilities exposed during recent crises—such as supply chain inefficiencies, climate-induced disruptions, and emergent risks associated with new technologies and operational environments—and the inadequacies in current resilience frameworks. It also highlights the dual role of maritime transport and supply chains as contributors to and victims of global challenges, particularly concerning greenhouse gas emissions and climate change impacts.

Building on this foundation, the final chapter synthesises successful stories and practical insights from preceding chapters, offering actionable strategies and the integral resilience model as a guiding framework. This model addresses the gaps in resilience planning identified in Chap. 1 by integrating adaptive capacities, collaborative networks, technological innovation and robust governance. Doing so aligns with the book's call to move beyond siloed approaches and adopt a holistic perspective that considers the complex interrelations among stakeholders, systems and external factors.

The roadmap provided in this chapter ensures that stakeholders—including port authorities, logistics operators, policymakers, researchers, and industry practitioners—are equipped to address emerging challenges proactively. It offers tools for planning and preparing for disruptions, absorbing and recovering from them, and adapting to ensure long-term sustainability and operational stability. In essence, this chapter reinforces the argument introduced in Chap. 1, which is that resilience is not merely about responding to crises, but about embedding a culture of adaptability, innovation and collaboration across maritime and supply chain networks.

By implementing the best practices and guidelines outlined herein, stakeholders can sustain trade flows, enhance economic stability, and contribute to global resilience. This comprehensive roadmap for advancing resilience in maritime systems underscores the interconnected nature of modern supply chains and the urgent need for coordinated, forward-thinking strategies to navigate an uncertain future.

References

Datta, B., & Satghare, H. R. (2024). *Global cargo industry: Resilience of Asia-Pacific shipping industries: Resilience of Asia-Pacific shipping industries*. IGI Global.

De Alwis, N., & Nam, H.-S. (2024). A way towards port automation: challenges and implications. *WMU Journal of Maritime Affairs*, 1–23.

Kin, B., Buldeo Rai, H., Dablanc, L., & Quak, H. (2024). Integrating logistics into urban planning: Best practices from Paris and Rotterdam. *European Planning Studies, 32*, 24–44.

Kok, B., Eski, Y., & Boelens, M. (2024). Port in an E-storm: Exploring awareness of state-sponsored cyber-threats and-attacks in the port of Rotterdam and North Sea Canal area. *European Journal of Policing Studies, 7*, 189–214.

Liu, Q., Yang, Y., Ng, A. K., & Jiang, C. (2023). An analysis on the resilience of the European port network. *Transportation Research Part A: Policy and Practice, 175*, 103778.

Mulopulos, S. (2023). *Supply chain resilience policy: Theory, practice, and action*. Fulbright New Zealand.

Taneja, P., Van Rhede Van Der Kloot, G., & Van Koningsveld, M. (2021). Sustainability performance of port infrastructure—A case study of a quay wall. *Sustainability, 13*, 11932.

Wilber, K. (2005). Introduction to integral theory and practice. *AQAL: Journal of Integral Theory and Practice, 1*, 2–38.

Index

Y.-y. Lau et al., *Maritime Transport and Supply Chain Resilience*,
https://doi.org/10.1007/978-3-032-07566-6

MIX
Papier aus verantwortungsvollen Quellen
Paper from responsible sources
FSC® C105338

FSC
www.fsc.org

If you have any concerns about our products,
you can contact us on
ProductSafety@springernature.com

In case Publisher is established outside the EU,
the EU authorized representative is:
Springer Nature Customer Service Center GmbH
Europaplatz 3, 69115 Heidelberg, Germany

Printed by Libri Plureos GmbH
in Hamburg, Germany